Advancing the U.S. Air Force's Force-Development Initiative

S. Craig Moore, Marygail K. Brauner

Prepared for the United States Air Force

Approved for public release; distribution unlimited

The research described in this report was sponsored by the United States Air Force under Contract FA7014-06-C-0001. Further information may be obtained from the Strategic Planning Division, Directorate of Plans, Hq USAF.

Library of Congress Cataloging-in-Publication Data

Moore, S. Craig, 1946–
Advancing the U.S. Air Force's force-development initiative / S. Craig Moore, Marygail K. Brauner.
p. cm.
Includes bibliographical references.
ISBN-13: 978-0-8330-4012-1 (pbk. : alk. paper)
1. United States. Air Force—Officers. 2. United States. Air Force—Occupational specialtiesl. 3. United States. Air Force—Personnel management. I. Brauner, Marygail K., 1947– II. Title.

UG793.M63 2007
358.4'130973—dc22

2007011005

Published 2007 by the RAND Corporation
1776 Main Street, P.O. Box 2138, Santa Monica, CA 90407-2138
1200 South Hayes Street, Arlington, VA 22202-5050
4570 Fifth Avenue, Suite 600, Pittsburgh, PA 15213-2665
RAND URL: http://www.rand.org/
To order RAND documents or to obtain additional information, contact
Distribution Services: Telephone: (310) 451-7002;
Fax: (310) 451-6915; Email: order@rand.org

Preface

The Air Force's force-development initiative evolved from research that the RAND Corporation began in the late 1990s, aiming to improve the development of Air Force senior leaders. This monograph summarizes force development's history, recommends ways to advance the initiative, and suggests areas that need senior leaders' attention and decisionmaking. In it we respond to discussions during 2005 with Maj. Gen. Tony Przbyslawski, Commander of the Air Force Personnel Center. In current force-development parlance, we concentrate on the development of "occupational competencies" that can be managed substantially and more definitively using officer assignments, in contrast to "institutional competencies" (such as negotiating, visioning, fostering diversity, listening actively, and demonstrating ethical leadership) that are less job-specific and that the Air Force is working to inculcate largely through education and training.

We wrote the monograph within RAND Project AIR FORCE's Manpower, Personnel, and Training Program under a project named "Methods and Prototypes for Air Force Force Development," which was sponsored by the Air Force Deputy Chief of Staff for Manpower and Personnel. This work's antecedents were projects sponsored by the Air Force General Officer Matters Office (later the Air Force Senior Leader Management Office), the Developing Aerospace Leaders Project Office, and the Air Force Space Command.

This document should be of value across the Air Force manpower and personnel community and perhaps even more to career-field managers and development teams in other functional areas. It aims to help members of the Force Management and Development Council (FMDC) (formerly the Force Development Council) adopt a common vision for the future of occupational force development and to promote consistent advancements within and across their functional areas. We are concerned that the force-development initiative eventually could come to little unless more Air Force leaders engage positively, vigorously, and definitively to advance its implementation.

Other published RAND research in this area includes the development and application of data and methods for targeting the occupational skills needed in the Air Force's senior military and civilian executives (at the level of military general officers and the civilian Senior Executive Service, see Robbert et al., 2004) and improving the utilization and development of Air Force space and missile officers at lower grades (regarding officers from lieutenant through colonel, see Vernez et al., 2006). Additional related forthcoming work will identify the occupational skills needed in the Air Force's colonel workforce and, paralleling the research that addressed space and missile officers, will identify improved career paths for Air Force rated officers (pilots, navigators, and air battle managers) and intelligence officers. Similar RAND research has addressed the backgrounds needed in military and/or civilian executives in the U.S. Army, Navy, and Marine Corps and in the Federal Bureau of Investigation. In addition, RAND has studied the development of Army tactical leaders (see Leonard et al., 2006), the utilization and management of general and flag officers across the Department of Defense (see Harrell et al., 2004), and how to develop joint officers at lower grades (see Thie et al. , 2005).

RAND Project AIR FORCE

RAND Project AIR FORCE (PAF), a division of the RAND Corporation, is the U.S. Air Force's federally funded research and development center for studies and analyses. PAF provides the Air Force with independent analyses of policy alternatives affecting the development, employment, combat readiness, and support of current and future aerospace forces. Research is conducted in four programs: Aerospace Force

Development; Manpower, Personnel, and Training; Resource Management; and Strategy and Doctrine.

Additional information about PAF is available on our Web site at http://www.rand.org/paf.

Contents

Figures

Tables

Summary

A mismatch in the late 1990s between the qualifications needed for key general officer positions and the available candidates' backgrounds stimulated an extensive U.S. Air Force effort to improve the development of senior leaders. The Air Force needed to develop cohorts (individuals who enter the force in a year's time) of senior officers—generals and colonels—who have sufficient breadth for their current jobs and for positions they may need to fill in the future. In the past, most officers were managed within their career fields and were too narrowly specialized. The intent of our research was to understand the types of skills (or "competencies") that each officer position really needs and then to set targets for the numbers of those who should acquire those skills over their careers before they are promoted to colonel. The effort evolved into the current force-development initiative, which is managed by the Deputy Chief of Staff for Manpower and Personnel; advised by the FMDC, chaired by the Vice Chief of Staff; and substantially carried out by career-field managers, functional development teams, and the Air Force Personnel Center's officer assignment teams. The initiative concentrates first on the development of officers in grades below colonel. While parallel systems are forming for enlisted and civilian personnel and the reserve components, our monograph addresses the system for officers in the active component.

The force-development system should aim to develop enough officers with specified backgrounds so that multiple qualified candidates will be available for each opening. Beyond the occupational specialties (primary skills) where they substantially "grew up," most general officers ideally would have significant experience in a secondary or paired occupation or skill, preferably with corresponding education or training. For example, a bomber pilot with a paired skill in international political-military affairs would be regarded as properly qualified for nearly twice as many general officer positions as one lacking a paired skill. It usually takes deliberate development for an officer to gain experience outside his or her primary specialty. Subsequent analysis extended the analysis of primary and secondary skills for general officers and formulated targets for paired skills when officers are promoted to colonel—targets that the development teams began using during 2005 to guide the mixes of developmental vectors selected for officers at lower grades in their career fields. For example, at least 5 to 6 percent (but preferably about 12 percent) of new mobility colonels ("rated" officers who grew up as airlift or tanker pilots or navigators) would have secondary skills in planning and programming, and about the same share would have secondary skills in acquisition or financial management (see p. 18). Factors like job sequencing (e.g., some jobs are appropriate first jobs for colonels while others require senior colonels), ill-shaped job pyramids (e.g., some skill pairs are needed for senior jobs only), and the need for selectivity (multiple qualified candidates should be available when openings occur) imply that substantially more officers than positions need paired skills. Flow analysis found that, overall, at least 31 percent, and preferably about 58 percent, of new line colonels should have secondary skills, even though in fiscal year 2002 only 23 percent of about 2,800 line colonel positions needed secondary skills. Naturally, the targeted secondary skills and percentages differ across career fields (see p. 16).

A four-step approach can create notably more-specific developmental targets for officers in grades below general officer within a particular career field: (1) identify and prioritize the types of experience, education, and training that should precede each category of job (identify the demand, at least for the jobs in the field grades—major, lieutenant colonel, and colonel), now and in the future; (2) ascertain the backgrounds that officers have accumulated (assess the supply); (3) compare supply with demand (gap analysis); and (4) plan ways to close the gaps. (See pp. 20–26.) We have demonstrated the approach for space and missile operations officers (the 13S career field) and are using it now with (and for) the "rated" (11X [pilot], 12X [navigator], and 13B [air battle manager]) and intelligence (14N) career fields. When coupled with careful management of officers' assignments and schooling, the approach promises far more complete fulfillment of positions' needs and far greater use of officers' backgrounds than are currently available (see Figure 4, p. 25).

Key Findings

Multiple Skills Required

Analysis of both general officer and colonel-level (O-6) positions shows that many positions need pairs of skills—primary and secondary (see pp. 2–3, 6–7, 11). That is largely why the Air Force instituted the FMDC and all the associated procedures, to ensure that enough officers get experience in multiple specialties and to counter the tendency toward overspecialization.

Assessing Skill Requirements

Systematically identifying (and periodically updating) positions' requirements for both primary and secondary skills is essential. While it may seem daunting at the outset, it has proven feasible to identify such requirements and to get them accepted across the Air Force, as RAND's work addressing the military and civilian executive forces demonstrates, for example. We recommend using expert panels rather than surveys of job incumbents to identify and update most skill requirements, especially for the field grades (see pp. 12, 22).

Important Implications of Multiple-Skill Requirements

Multiple-skill requirements have extensive ramifications, as found through flow analysis. Even if a minority of positions demands multiple skills, to meet those demands a majority of the officer cohort may need to have multiple skills. The inventory must include many multi-skilled officers, in spite of officers' natural desire to stay within their primary specialty in each assignment (see Table 4, p. 16).

Keeping Officers on the Right Path

The Air Force needs continuing management to ensure that officers, at least in the aggregate, are following the right pathways. In particular,

- The Air Force needs agreed-upon mechanisms for tracking skills, especially to answer the question: How much is enough? This remains an open question, since originally it was believed that two tours were needed in a secondary skill, but since then it has become policy to grant a "developmental [skill] identifier" after just one year of experience (see p. 9). Having just one year's experience in an occupational area, perhaps even early in their careers, probably falls far short of making officers viable candidates for leadership positions in those areas once they become colonels or generals.
- The development system needs to manage career fields so that enough officers have the targeted skill pairs by the time of a cohort's promotion to O-6. However, a paired skill should not be essential for promotion to colonel (see p. 19).

Next Steps

We recommend that the Air Force make its formal instruction about officer force development (AFI 36-2640) more specific regarding the need for paired skills at senior levels and regarding the need to systematically plan and manage development for far larger numbers of midcareer positions. We believe the FMDC and its members should play important roles in shaping, advocating, coordinating, and monitoring how the functional communities (e.g., operations, intelligence, logistics, personnel, or acquisition) execute their force-development responsibilities.

The force-development community should take the following steps:

- Establish standards for earning paired skills, more demanding than earning developmental identifiers (see p. 9).

- Update and extend the database of backgrounds needed for colonel positions (see p. 12).
- Clarify that earning a paired skill is not essential for promotion to colonel (see p. 19).
- Set development teams on course to create measurable developmental targets for grades below general officer (see pp. 19–27).
- Create and monitor measures tracking the development of officers promoted to each grade, serving in command jobs, attending developmental courses in residence, or holding other key assignments (see pp. 27–28).
- Find ways to use or enhance data systems to (1) consistently register jobs' needs for prior experience, education, and training; (2) track individual officers' accumulating portfolios of experience, education, and training; and (3) help recommend and make assignments whose demands officers meet, that use the officers' backgrounds, that help manage career fields properly, and that match members' preferences insofar as possible (see p. 28–29).
- Improve force planning and management so that career fields' numerical strengths align more consistently with requirements and leave room for deliberate professional development (see p. 29).

Beyond yielding information that the Deputy Chief of Staff for Manpower and Personnel (AF/A1) and the FMDC can use in deciding resource allocations and making system adjustments, these steps should help the wide range of force-development players to develop consistent, efficient, and effective plans and means for improving the development of officers in their career fields. The steps are also likely to provide insights and mechanisms that will be valuable more widely as the Air Force extends and enhances force development to address the enlisted and civilian forces and the reserve components.

Acknowledgments

We are grateful to numerous Air Force officers and civilians for meeting with members of our research team and discussing candidly their perceptions of the evolving force-development system's strengths and weaknesses. Among our colleagues at RAND, Michael Thirtle was instrumental in arranging, conducting, and summarizing those interviews; Robert Guffey improved the monograph's organization and presentation; and Ray Conley, Brent Thomas, and Manuel Carrillo played key roles in analyzing colonel jobs and personnel flows in the work that helped generate skill-pairing targets for the development teams. In the Air Force Senior Leader Management Office, Gwen Rutherford helped release for review by career-field managers and development teams the colonel positions' requirements from fiscal year 2002, and Maj. Dan Gregg and Capt. Paul Emslie helped formulate the subsequent personnel flow analysis. Along with the Air Staff's John Park, Maj. Eric Johnson, and Maj. Todd Sriver and with the Air Force Personnel Center's Col. Scott Davis, Jerry Ball, Lt. Col. Josh Jose, Lt. Col. Harold Huguley, Maj. Rob Ramos, Maj. Jim DeHaan, Maj. Matt Santoni, Capt. Jeremy Sherette, and 1st Lt. Damon Richardson, they also helped frame and revise the analytic results that became the skill-pairing targets. Lt. Gen. Roger Brady chartered and Maj. Gen. Glenn Spears and John Park chaired meetings of a working group of career-field managers who reviewed that cooperative effort's methodology and preliminary results, evaluating the targets' potential utility and helping formulate their presentation to others. The working group included Col. Jeff Fraser for the rated community; Col. Wayne Hudson for space and

Force identify the need for more-deliberate force development and establish targets to guide the development of future senior leaders, supporting the force-development initiative through its early years. Later, RAND Corporation assessed and developed methods to improve the finer-grained development and sustainment of officer workforces in specific career fields. From those perspectives, RAND offers this brief monograph as food for corporate thought in the Air Force, intending it principally for senior leaders, including the Air Force Deputy Chief of Staff for Manpower and Personnel (AF/A1) and staff, plus members of the Force Management and Development Council (FMDC) (formerly the Force Development Council), who are charged with guiding force development's implementation.[2] We believe the FMDC should inspire and help lead the initiative, guiding it toward an institutional, strategic perspective that cuts across career fields, ensuring that individual career fields develop force-wide goals consistent with that perspective, and tracking progress toward meeting both cross-cutting and career field–specific goals.

Paralleling the Air Force's initial emphasis, this monograph concentrates on occupational force development for the active component's officer force. It seems worth remembering the late Gen. Robert Dixon's observation as advisor to then–Chief of Staff of the Air Force General Ryan: "Transforming officer development is more important to the Air Force's future than acquiring the F-22 and the Joint Strike Fighter. It will be harder to do, and there's greater risk of failure" (quote from the Senior Leader Kickoff meeting, 1999).

The key difficulty was that most Air Force officers grew mainly within narrow, primary occupational areas, such as fighters, intelligence, or maintenance, and became well-qualified for relatively few senior positions that require those specialized skills. Officers with a paired (or secondary) skill would be viable candidates for many more senior positions. For example, bomber pilots and navigators without a paired skill were best qualified for only about 6 percent of 2005's general officer jobs, but a paired skill in acquisition management would

[2] U.S. Air Force (2004) describes the force-development program and spells out the FMDC's and others' responsibilities.

qualify them for another 3 percent of the jobs, a paired skill in international political-military affairs for another 5 percent, a paired skill in planning and programming for another 7 percent, and a paired skill in air power employment for another 15 percent.[3] Positions for colonels exhibit similar requirements, although relatively fewer of them require paired skills. Our research set out to understand the types of skills (or "competencies") that each senior position really requires in order to target how many officers should acquire those skills before they are promoted to colonel.

This monograph reviews steps that have shaped the force-development initiative, including some missteps that illustrate how easily the initiative can get off track. Then it sketches the targeting of occupational skills—specifically, paired primary and secondary occupations—needed in future senior Air Force general officers and colonels, so far in all except the medical, legal, and chaplain specialties. Then it describes and illustrates an approach for planning much finer-grained development of officers for the field grades—major, lieutenant colonel, and colonel. It closes with recommendations for next steps for the force-development initiative.

History, Including Some Missteps

The force-development initiative began with a new way of thinking about the occupational competencies required for general officer and SES positions. As early as 1998, the Air Force identified individual general officer (and later SES) positions' needs for occupational competencies, such as fighter, bomber, intelligence, maintenance, planning and programming, aerospace power employment, and information operations. Air Force leaders also identified cross-functional competencies that all senior officers should have, such as management, analysis, and communication skills, although with different degrees of emphasis for

[3] As Robbert et al. (2004) explain, the analysis considers generals qualified for a job if either their primary or, less desirably, their secondary skill matches the job's required primary skill. This would qualify bomber pilots or navigators with a secondary skill in, say, acquisition management for nearly 47 percent of the total general officer positions.

different jobs.[4] The occupational requirements fed into an analysis of potential promotion and utilization patterns to help increase the likelihood that more incoming military and civilian executives would have the combinations of skills needed for senior jobs.[5] See Robbert et al. (2004).[6]

Further steps were taken in 2001, when, with the Corona's endorsement,[7] General Ryan and Secretary of the Air Force Whitten Peters established the Developing Aerospace Leaders (DAL) initiative and the DAL Program Office to plan deliberate ways of developing future leaders. DAL considered the officer, enlisted, active, guard, reserve, and civilian workforces, concentrating first on the officer force. It helped mature the idea of paired occupational competencies, recommended that standards be established for certifying officers' primary and secondary occupational proficiencies, proposed realigning professional military education to help support development (renaming it developmental education), and recommended consolidating existing functional managers' career-field management activities into fewer, larger "core specialty-management" offices.[8] Also, at a Corona conference in

---

[4] The "cross-functional competencies" morphed into today's "institutional competencies" that the Air Force is pursuing outside of and across occupational channels.

[5] In this document the terms *needs*, *requirements*, *necessary skills and competencies*, *demand*, and *required background* mean the same thing: the previous experience, education, and training that Air Force members should bring to their jobs. Naturally, different jobs call for different backgrounds.

[6] Although the fact is not documented in that technical report, personnel records for officers promoted to general officer and those who are competitive for promotion to general officer showed that few had developed expertise beyond their own occupational "stovepipes"—i.e., most were narrowly rather than broadly skilled and, consequently, were well prepared for relatively few senior leadership positions.

[7] The Air Force's four-star generals assemble three times each year in Corona meetings.

[8] The DAL Project Office recommended core specialty managers for a dozen areas: air combat operations; air mobility operations; space operations; information warfare operations; command, control, communications, computers, intelligence, surveillance, and reconnaissance (C4ISR) operations; special operations; political-military strategy; systems acquisition; logistics operations; maintenance; installation operations; and human resources operations. Lieutenants and captains would develop as traditional specialists (e.g., as fighter pilots, intelligence officers, aircraft maintenance officers, personnel specialists, or acquisi-

2003, Chief of Staff of the Air Force John Jumper and Secretary of the Air Force James Roche opted to simplify and operationalize many of these basic ideas under the force-development moniker and to

- use the existing functional management framework instead of consolidating into core specialties
- establish the FMDC to oversee force-development policies and processes
- create functional development teams to guide the career paths of the officers in their career fields
- reorganize offices at the Air Staff and the Air Force Personnel Center (AFPC) to support the FMDC and development teams.

To help the development teams guide officers into appropriately paired occupations, the Air Force Senior Leader Management Office (AFSLMO) consulted with the functional managers and, in 2003, issued the occupational skill requirements listed in Table 1. Its major headings, such as Logistics, Fighter, and Intelligence, name primary occupations where leading officers should have spent the bulk of their careers, and the subordinate lists name the occupations or paired skills where they should develop a secondary competency. A leader with a primary background in space might have a paired skill in acquisition, communications, or aerospace power employment, for example.

The next step was to target rough numbers of new graduates per year from intermediate developmental education (for majors) who

---

tion specialists); majors and lieutenant colonels would develop as "core specialists" (leading or managing the integration of multiple specialties' contributions within one of the 12 core areas); and some officers would develop further as "aerospace specialists" (broadening into one of 14 deliberately paired application areas: joint operations, aerospace operations, air combat, air mobility, space, information warfare, C4ISR, plans and programs, political-military strategy, acquisition, logistics and maintenance, support operations, institutional sustainment, and education). Enough colonels and generals were to become aerospace specialists to create an adequate "bench" from which to fill senior jobs for operations, information, force support, and materiel "transformational leaders" and ultimately for the most senior jobs for aerospace employment, aerospace component commanders, joint employment, and joint leadership. During the DAL period, the family of nonoccupational competencies, earlier labeled "cross-cutting" and today called "institutional," were called "universal" and then "enduring" competencies.

Table 1
AFSLMO-Issued Skill Requirements

Primary Occupation	Paired Secondary Occupations
Fighter	Air power employment Political-military Logistics Plans and programs Acquisition Information operations Space Education and training
Bomber	Air power employment Acquisition Space Political-military Logistics Plans and programs Information operations Education and training
C2ISR-rated[a]	Air power employment Information operations Space Logistics Plans and programs Acquisition Political-military Education and training
Mobility (tanker and/or airlift)	Air power employment Mobility operations Acquisition Space Logistics Plans and programs Political-military Information operations Education and training
Special operations	Air power employment Space Plans and programs Logistics Acquisition Political-military Information operations Education and training
Space	Acquisition Communications Aerospace power employment Intelligence Plans and programs Information operations Political-military Education and training

RAND *MG545-TABLE1*

Table 1—Continued

Primary Occupation	Paired Secondary Occupations
Intelligence	Political-military Information operations Aerospace power employment Space Plans and programs Personnel Education and training
Maintenance	Logistics Financial management Aerospace power employment Plans and programs Acquisition Political-military Space Education and training
Logistics	Maintenance Contracting Financial management Aerospace power employment Plans and programs Political-military Acquisition Education and training
Communications and information systems	Information operations Intelligence Plans and programs Aerospace power employment Space Education and training C2ISR
Acquisition management	Maintenance Space Information operations Aerospace power employment Plans and programs Political-military Education and training
Other occupations	Information operations Space Acquisition Aerospace power employment Political-military Plans and programs Mobility operations Financial management Personnel and manpower Intelligence Education and training

[a]Rated officers hold aeronautical ratings as pilots, navigators (weapon system officers), or air battle managers. Rated occupations in this table include fighter; bomber; C2ISR-rated; mobility; and special operations.

RAND *MG545-TABLE1 (cont.)*

of-experience profile, plus each grade's authorized and assigned personnel; (2) the fields in which members have secondary or paired skills; (3) how various factors have influenced promotion rates; and (4) promotion rates compared with those for the force as a whole. However, these assessments lacked requirements-based targets for paired skills and other qualifications that are needed in order to understand how officers should be moving through their career fields.

The Air Force has begun taking a requirements-based approach to force development, aiming to ensure that enough officers gain the right kinds of skills in time to perform the jobs at colonel and general officer level, but the Air Force is not yet systematically or comprehensively planning and managing the development of officers to perform the jobs at the middle levels within individual career fields. The next three sections describe how the targets were derived for senior leaders' skill pairings, outline a complementary approach that can guide the development of mid-level officers in the various career fields, and recommend steps the Air Force can take toward more complete implementation.

Developing Future Senior Leaders (General Officers and Colonels)

An important part of the force-development initiative is the establishment of inflow goals, which identify the combinations of skills needed in the force that will fill senior positions. These inflow goals help development teams, career-field managers, and AFPC grow enough officers with the right combinations of skills to fill the senior jobs.

The methods for analyzing general officer and SES flows and targeting inflows at that level are documented in Robbert et al., 2004, as already noted. Subsequent efforts focused on developing inflow goals for new colonels, establishing more immediate development targets for officers through lower grades (from lieutenant through lieutenant colonel), when the development teams, career-field managers, and AFPC directly affect officers' development. A part of this effort was broadened to include civilian GS-15 jobs, the source of most new SES members. AFSLMO, assisted by RAND, took a three-step approach:

1. Identify the skills needed to fill colonel positions.
2. Develop a flow-analysis model to help translate job requirements into goals for the annual inflow of new colonels.
3. Use analytic results to target skill pairings.

Identifying the Skills Needed to Fill Colonel Positions

In 2002, panels of functional experts had identified the primary and secondary occupations needed for each of about 2,800 of fiscal year 2002's colonel jobs, excluding only the medical, legal, and chaplain career fields.[11] Analysis of the data found that, collectively, the specifications were relatively flexible about the primary and secondary skills required for colonel positions:

- About 20 percent of the colonel jobs could accept officers from any primary career field, and another 40 percent were open to more than one career field.
- Requirements for paired skills were even more flexible. About 77 percent of the jobs were judged to need no secondary occupation at all, and about a third of the others could accept more than one secondary occupation.
- Many requirements did not match the development teams' career fields. Some were broader—e.g., any rated specialty, any operations specialty, and either acquisition or logistics. And some were narrower—e.g., fighter, bomber, missile, and satellite command and control.
- Authorized specialties (Air Force specialty codes) often were too restrictive—e.g., only about 12 percent of the 126 jobs authorized for fighter pilots (11F) could accept only fighter pilots, and only about 22 percent of the 195 jobs authorized for acquisition managers (63A) could accept only acquisition managers.

[11] Note that functional managers, major commands, and AFSLMO reviewed and refined the expert panels' results. Moreover, there is unpublished RAND research by S. Craig Moore and Brent E. Thomas on targeting the occupational skill pairings needed in new Air Force colonels.

On a nearly emergency basis, AFPC aggregated some of these requirements and fair-shared others to produce interim skill-pairing targets for fiscal year 2005's summer meetings of the development teams. The process occasionally yielded some odd pairings—e.g., when a less-specific need parsed into a requirement for civil engineers with a paired skill in manpower and personnel. Most development teams never received those targets, and only one or two used them. The summer meetings concentrated instead on recommending which officers should attend schools.

In parallel with the analysis of fiscal year 2002 data, AFSLMO launched an effort to survey colonels about the backgrounds needed for their jobs; update the primary and secondary skill requirements; address additional requirements for experience, education, and training; and extend the scope to cover GS-15 positions. AFSLMO distributed the survey in June 2005, but only about 40 percent of recipients responded. *If its results are to become useful, many more survey responses must be obtained.* Then, functional experts would need to review the results and fill in all missing data.

Development planning needs a full picture of the range and mix of requirements anticipated for the future. Our experience with this and other large-scale efforts to ascertain job-specific competency requirements suggests that *panels of subject-matter experts rather than surveys of incumbents offer a more manageable and reliable approach.*

Lessons from a New Flow-Analysis Model

RAND and AFSLMO analysts developed a new flow-analysis model to help translate job requirements into goals for the annual inflow of new colonels. The model showed that, although many colonel jobs have a fair amount of flexibility in terms of the primary and secondary skills they require, the mix of incoming colonels and the paths they must follow to fill those jobs are much less flexible. There are several reasons:

- *Sequencing.* Some jobs are appropriate as first jobs for colonels and some as second jobs, but some need senior colonels.
- *Occupational pyramids.* Some skill pairings are needed for senior jobs only, for example.

- *Preparatory roles.* Some jobs are important for preparing and/or testing colonels as candidates for promotion to general officer.
- *Selectivity.* Several qualified candidates should be available when openings occur, so that a choice of good candidates can be offered for the selection process.
- *Progression.* Nominal continuation rates and job tenures affect availability for selection in future openings.
- *Outflows.* The colonel force must yield enough competitive candidates with the skill pairings needed to feed into and sustain the general officer force.

Users can guide the new model by targeting minimum levels of selectivity (the number of candidates from whom one can choose when filling a position vacancy), how precisely the jobs' experience designations must be met, and how many more general officer candidates must be made available than can actually become general officers, for example. The model identifies flows that would minimize the number of incoming colonels with paired skills, maximize flexibility in the occupational mix of new colonels, maximize its congruity with a targeted mix of primary occupations, and maximize similarity in the fractions of incoming cohorts from different occupations who would possess paired skills.

RAND ran the model with the skill-pairing requirements that the expert panels had identified for fiscal year 2002's line colonel jobs (i.e., excluding the medical, legal, and chaplain corps). Figure 1 illustrates the results. The bar on the left shows the amount of flexibility in the job requirements, subdivided into jobs that require specific primary and secondary skills (3 percent), jobs that require a specific primary skill and allow flexibility in the secondary skill (32 percent), jobs that require a specific secondary skill and allow flexibility in the primary skill (12 percent), and those that allow flexibility in both the primary and secondary skills (53 percent). The two personnel inventory bars on the right reflect two variations on the policy goals for the actual inventory to meet the job requirements. The middle bar is a "looser" specification, providing marginally acceptable flows; the right-hand bar is

Figure 1
Flow Analysis Reduces but Does Not Eliminate Flexibility

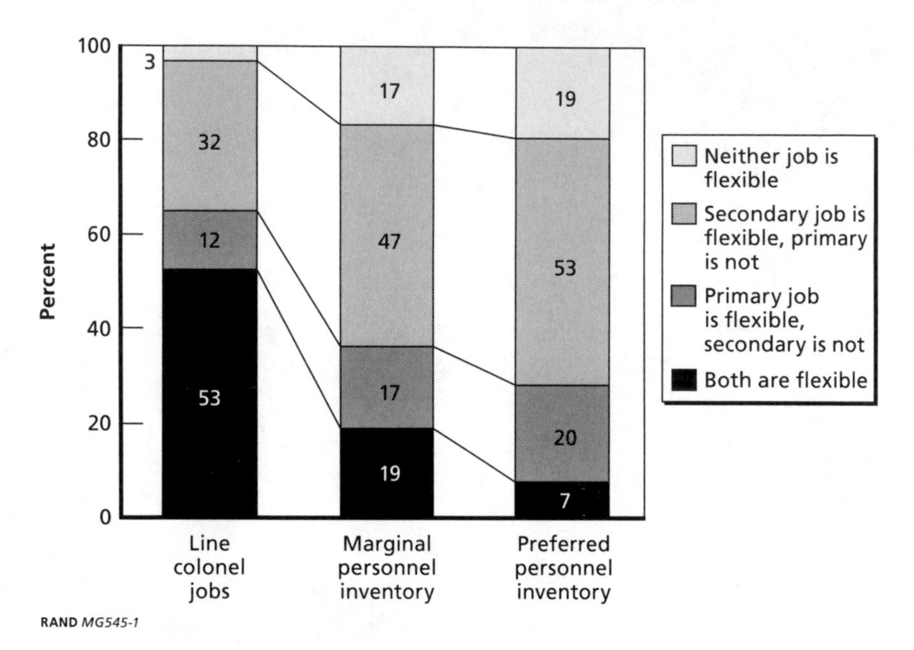

RAND *MG545-1*

"tighter," providing preferable flows. The case labeled "marginal personnel inventory" seeks at least two qualified candidates per opening and would fill at least half of the jobs that require them with senior-level colonels, for example, while the case labeled "Preferred personnel inventory" seeks at least three qualified candidates per opening and would fill at least 90 percent of the jobs that require them with senior-level colonels. The higher the selectivity target and the more nearly the experience and other policy targets must be met, the less flexibility remains in the targeted personnel inventory.

Table 2 illustrates how the amount of flexibility in the inventory of colonels decreases as the colonels better match the skills required for their jobs. Fifty-one of the fiscal year 2002 positions could accept colonels with either fighter or bomber backgrounds as the primary occupation. In the marginal case, the model recommended using fighter colonels for 56 percent of those positions and bomber colonels for 13 percent; the remaining 31 percent could be either fighter or bomber

Table 2
Inventories Recommended for Jobs Accepting Either Fighter or Bomber Experience as the Primary Occupation (percent)

Primary Occupation	Marginal Case	Preferred Case
Fighter	56	73
Bomber	13	26
Flexible	31	1

colonels. In the preferred case, it recommended using fighter colonels for 73 percent and bomber colonels for 26 percent of the inventory, leaving only 1 percent of the inventory flexible to accept either fighter or bomber colonels. As noted above, these numbers reflect the demands of sequencing, occupational pyramids, preparatory roles, selectivity, progression, and outflows.

The analysis also revealed reduced flexibility among secondary occupations. Table 3 matches the modeled inventory against the 50

Table 3
Inventories Recommended for Jobs That Require Intelligence as the Primary Occupation and Are Flexible About the Secondary Occupation (percent)

Secondary Occupation	Marginal Case	Preferred Case
Education and training	3	23
Plans and programs	0	20
Space and missile operations	10	19
Foreign area specialist	5	15
Aerospace power employment	0	9
Other	2	13
Flexible	80	0

NOTE: Numbers are rounded and may not add to 100.

fiscal year 2002 colonel positions that required intelligence as the primary occupation and that were flexible about the secondary occupation (part of the category of jobs shown second from the top in Figure 1). The flow analysis found it necessary to fill those positions 20 percent of the time using colonels with specific paired occupations—such as education and training, plans and programs, space and missile operations, and foreign area specialist—in the marginal case, and 100 percent of the time in the preferred case.

Finally, flow analysis shows that *far more officers than jobs need paired skills.* Based on the analysis described above, 23 percent of colonel positions required paired skills; to fill those positions, 31 percent of incoming colonels need paired skills in the marginal case and 58 percent in the preferred case. Table 4 illustrates this pattern for several groups of career fields. For example, while the experts said that only about 29 percent of fiscal year 2002's colonel positions that needed acquisition and finance officers also needed paired skills, flow analysis found that at least 38 percent of new colonels from acquisition and finance should have acquired paired skills in the marginal case and 68 percent in the preferred case.

Table 4
Far More Officers Than Positions Need Paired Skills (percent)

Occupational Group	Proportion of Fiscal Year 2002's Positions Needing Paired Skills	New Colonels Needing Paired Skills	
		Marginal Case	Preferred Case
Rated	21	24	53
Nonrated operations	48	62	93
Logistics	11	37	56
Support and special investigators	7	21	36
Acquisition and finance	29	38	68
More than one group	28		
Total	23	31	58

Using Analytic Results to Target Skill Pairings

The Air Force has begun to use results from the analytic approach described above. When it became clear in 2005 that the new survey could not authoritatively update the skill requirements for the colonel positions soon enough, the Air Staff and AFPC directed a team of their analysts to work with RAND and use flow analysis and the requirements identified in 2002 to recommend skill-pairing targets that the development teams could consider as guidance for the fall meetings on vectoring. The group met several times to explore and understand the methodology, examined alternative assumptions and priorities that affect its calculations, and tried different ways of organizing and packaging its results.[12] Figure 2 illustrates targets for the mobility career field (airlift and tanker pilots and navigators in the mobility Air Force). It shows the minimum percentage of new mobility colonels who should have a particular paired skill, expressed as a range from marginal to preferable. Such visual displays help development teams make vectoring decisions that should result in better inflows of qualified colonels.

It is worth knowing that, as the executive agent for the FMDC, Lt. Gen. Roger Brady, the Air Force Deputy Chief of Staff for Manpower and Personnel (AF/A1) asked several functional managers to appoint members to an FMDC skill-pairing working group to critically review those results and recommend ways to improve them before distributing them more widely and authoritatively for development teams' implementation. Some working group members questioned a few of the pairings but usually were satisfied when they were able to trace targets back to the jobs that justified them. The greatest concerns arose because (a) some colonel jobs had been eliminated and others created since 2002; (b) a few position's required skill pairings were questionable; and (c) post-modeling allocations of remaining flexibility created a few

[12] The analysts judged it important to convey the skill-pairing targets as ranges, not as precise, definitive percentages. As Figure 3 suggests, each range represents a floor. Selectivity will be higher and good person-to-job matches will be more likely if more officers develop each paired skill. The closer the result is to the high end of each range (or even beyond the high end), the better for those purposes. On the other hand, it is inappropriate to go much higher if developing a paired skill displaces the development of important depth and expertise within an officer's primary career field.

or skill-pairing targets in order to plan the development of officers for jobs at the grades of major (O-4), lieutenant colonel (O-5), and colonel (O-6). Including colonel positions when planning both institutional, forcewide development and development for career fields' midlevel jobs should foster planning consistency and compatibility, even though different functional managers oversee the various career fields.

Because acceptable skill-pairing targets were lacking until recently, development teams naturally gravitated toward developing officers for success within their own career fields. While many criteria have not been formalized,[13] officers inevitably form impressions of successful career paths by observing their predecessors. Also, existing career-path guidance sketches nominal progression through organizational levels, education, and grades, as is illustrated in Figure 3 for aircraft maintenance officers.

Even though few development teams had provided officers with vectors for paired skills until the fall of 2005, most apparently perceived great value in meeting regularly to systematically assess and guide officers in the middle grades about desirable career vectors. The development teams began the important processes of reviewing officers' records, preferences, and career potential and recommending organizational and educational vectors to guide individuals' professional development. As noted, until quantitative goals for skill pairing emerged, the development teams understandably concentrated on individual officers.

RAND developed and demonstrated a four-step approach that the Air Force can use to establish well-justified targets for entire career fields. This approach grew out of work for the Air Force Space Command that assessed the assignment and utilization of space and missile operations officers (the 13S specialty) and the career field's sustainability (Vernez et al., 2006). Since completing the project in 2003, RAND

---

[13] Some career fields have more-definitive requirements than others, of course. For example, the rated career fields have formal "gate" and currency programs requiring fairly regular accumulation of flying experience, and the Defense Acquisition Workforce Improvement Act requires systematic, progressive qualification and certification for many positions in the acquisition career fields.

Figure 3
Aircraft Maintenance Careers

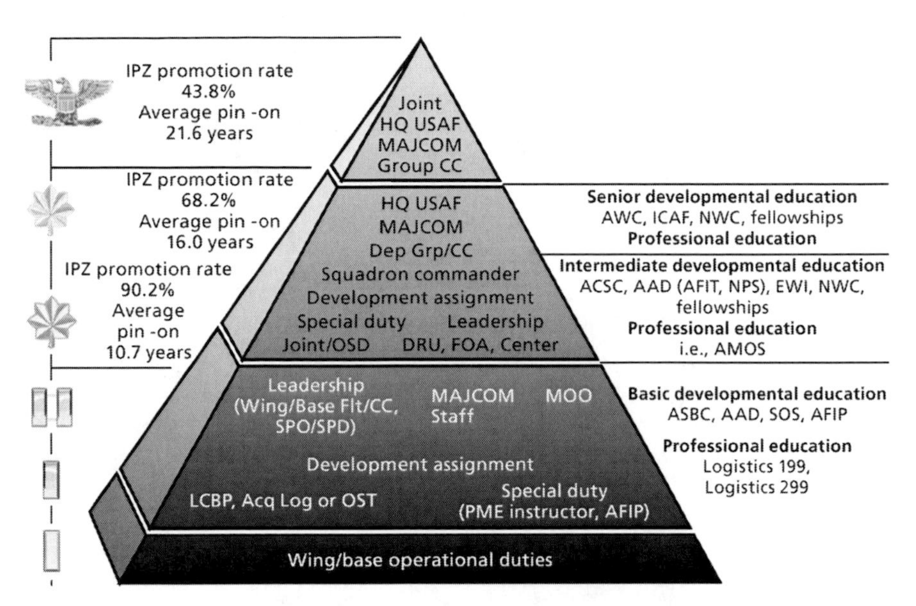

SOURCE: Air Force Personnel Center, "21A Aircraft Maintenance: Career Pyramid," Fall 2005. Online at http://ask.afpc.randolph.af.mil/main_content.asp?prods3=247&prods2=244&prods1=1&cats1=141&p_cats=141&p_faqid=6022 (Officer Force Development, Career Planning Diagrams & Utilization Charts, Non-Rated Ops) (as of August 18, 2006).

NOTES: The items listed inside the pyramid reflect categories of jobs that a maintenance officer may expect to hold at each stage of his career. The list on the right reflects developmental and professional education scheduled for each career phase. The list on the left reflects pay grades, nominal promotion rates, and the timing of each grade change, along with the corresponding insignia earned for each phase. From the bottom, the phases and associated pay grades are second lieutenant (O-1), first lieutenant (O-2), captain (O-3), major (O-4), lieutenant colonel (O-5), and colonel (O-6).

RAND *MG545-3*

has undertaken parallel research addressing the rated and intelligence officer forces. Because only the research on the space and missile officer force is complete, the following description of the four steps uses illustrations from the 13S career field.

Step 1: Identify the Demand

This step draws judgments from experts, primarily colonels, about the importance of dozens of elements of background as preparation for different jobs or groups of jobs in the career field. Experts rate each prior experience, education, and training element as follows:

- **Critical:** *Absolutely essential* to effective performance of the job. Without this experience, the position holder could not perform the job.
- **Important:** *Helpful, but not essential* to effective performance of the job. Without this experience, the position holder could still perform the job, although it would be considerably more difficult and time-consuming.
- **Useful:** *Good, but not necessary* to perform the job. Without this experience the position holder could perform the job but with occasional difficulty.
- **Not relevant:** This background is rarely or never helpful to an officer in this job.

In the 13S career field, experts rated an average of 5.0 (out of 70) elements as critical or important for O-4 jobs, 6.2 for O-5 jobs, and 10.4 for O-6 jobs.[14] Table 5 shows the shares of about 1,100 jobs at grades O-4 through O-6 for which the experts said some of the backgrounds were critical or important. For example, they rated prior functional experience in plans and programs as critical or important for

[14] Assembling this information about the jobs' demands for prior background involved four principal steps: (1) officers at the Air Force Space Command and the Air Staff identified elements of experience, education, and training that may be needed for one or more positions; (2) working separately, about 50 experts identified and prioritized prerequisite elements for the roughly 10 to 30 jobs under their purviews; (3) assignment officers at AFPC provided similar information for about one-third of the jobs not characterized in number (2); and (4) a team of eight 13S colonels met for two days and carefully reviewed and refined the prerequisites and priorities. In subsequent, parallel work that addressed rated and intelligence jobs, it was more efficient to simply convene concentrated, multiday workshops in which experts identified and prioritized the experience, education, and training needed for specific groups of jobs and discussed potential future changes.

Table 5
Proportion of Jobs Requiring Prior Experience and Education for Space and Missile Operations Officers
(percent)

Prior Background Needed (examples)	O-4	O-5	O-6
Mission operations experience			
Satellite command and control	13	11	20
Missile crew	25	17	20
Special experience			
Squadron operations officer	2	22	31
Contingency and war planner	10	10	20
Standards and evaluation examiner	30	29	34
Functional experience			
Plans and programs	18	24	43
Acquisition	10	19	29
Organizational experience			
Wing level	20	21	41
Headquarters Air Force Space Command	20	42	64
National Reconnaissance Office	10	15	20
Command experience			
Squadron	3	11	64
Group	N.A.	1	36
Education and training			
Engineering degree	8	5	16
Must hold authorized grade	55	64	83

NOTE: N.A. = not applicable.

18 percent of the 13S O-4 jobs, 24 percent of the O-5 jobs, and 43 percent of the O-6 jobs.

Looking to the future of this career field, RAND systematically estimated how demand would change (1) if prior experience in both acquisition and warfighting functions and organizations were important for all commander jobs and (2) if space systems were "weaponized" and some support activities were civilianized.

Step 2: Assess the Supply

This step carefully reviews officers' personnel records to discover which elements of experience, education, and training they have acquired. For example, considering the same 70 elements that may be needed for 13S O-4, O-5, and O-6 jobs, we found that 13S lieutenants had acquired an average of 1.9 elements, captains 4.8, majors 8.6, lieutenant colonels 11.0, and colonels 13.5. This step can also identify the career paths followed by current and past officers and whether retention and promotion patterns vary among different groups of officers.

Step 3: Compare Supply with Demand (Gap Analysis)

This step ascertains whether enough officers at each grade have each element of experience, education, and training needed. Do they have them in the right combinations? Did they bring the backgrounds needed for their current jobs?

For the 13S career field, although enough officers at each grade usually had each element of background and each combination of elements, gaps were often wide between an officer's background and the prior experience, education, and training needed for his or her job. These gaps are illustrated in the bars marked "actual" in Figure 4, which portray the average (over the jobs at each grade) number of experience categories (a) required for the job but not present in the incumbent, (b) required for the job and possessed by the incumbent, and (c) possessed by the incumbent but not required for the job. About half of a job's needs were not met, on average. Notably, for about 90 percent of the jobs above O-3 that needed an officer with certain experience, the jobholder lacked one or more of the needed types of experience. Moreover, about two-thirds of the assigned officer's accumulated background elements were not needed for the job, on average. Many assignments apparently had been made with insufficient regard for the job's needs and the officer's background.[15]

---

[15] Actual assignments fall short of optimized results for understandable reasons—e.g., compared with the data assembled for our research, assignment teams have less complete, less consistent data about officers' backgrounds and jobs' needs; cohort sizes and assignment

Figure 4
Optimized Development and Utilization Patterns Provide a Better Match Between the Needs of Positions and the Prior Experience of Candidates

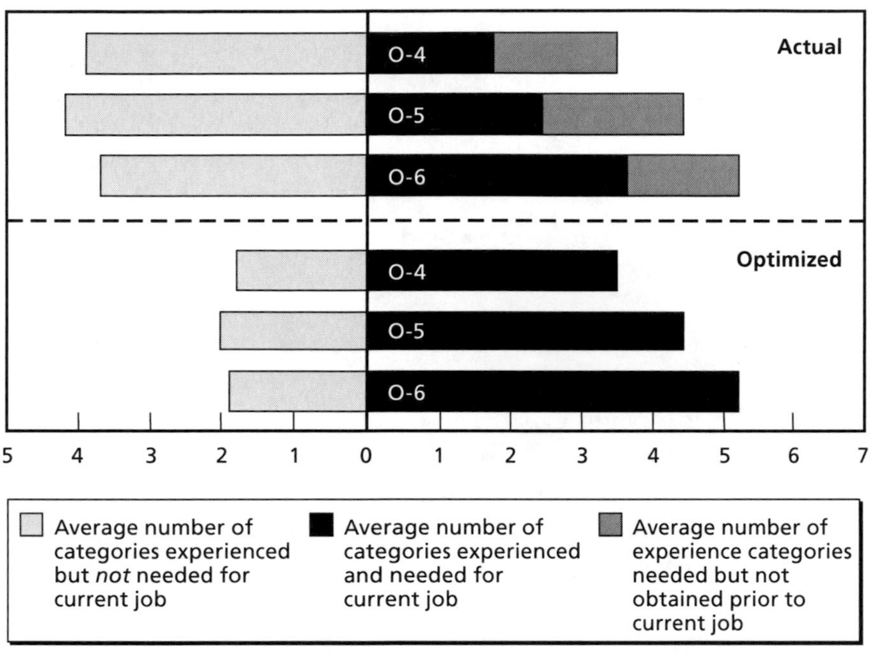

RAND *MG545-4*

Step 4: Plan Ways to Improve the Match Between Supply and Demand

What career paths would efficiently and systematically give enough officers the experience, education, and training needed for each group of jobs? How many officers should acquire different targeted mixes of backgrounds by each career point? Where should one look for officers with the background needed for a particular job and who would benefit from the experience the job gives? Having accumulated a given

---

guidance change over time; openings must be filled using only the officers available during an assignment cycle (many fewer than the full inventory); many "by-name" requests must be filled regardless of the job's or the officer's specific needs; and quantitative goals are lacking for using and developing officers' backgrounds. The force-development initiative aims to help assignment teams better see and meet jobs' and officers' needs.

combination of experience, education, and training, which kinds of jobs would best come next? RAND's work on space and missile officers shows how analysis can address such questions by modeling officers' sequences of assignments.

An initial model for the 13S force represented up to 13 career stages, from second lieutenant through colonel, with one to three jobs per grade. It recommended ways to develop and employ successive cohorts (the officers who enter the occupation in a year's time), matching jobs and (modeled) officers in ways that would best meet the jobs' demands and achieve policy-oriented objectives such as maximizing the depth or breadth of experience, establishing as many officers as possible on a "career track" by a certain point in their careers (distinguishing space, missile, and acquisition tracks), and preserving equal opportunity for advancement to higher grades. It awarded points each time an officer brought experience regarded as critical, important, or useful for his or her new job. To limit the model's size, staff at the Air Force Space Command consolidated most of the backgrounds considered in the demand, supply, and gap analyses into 12 broader categories of experience.

The bars marked "optimized" in Figure 4 illustrate the potential benefits of optimizing officers' career paths. Whereas the actual officers analyzed scored only 63 percent (based on the mismatches between backgrounds and requirements described under Step 3 above), the optimized flows would achieve 99.5 percent of the perfect score and would leave only a few majors lacking just one of their jobs' targeted types of experience. Aiming for depth of experience, the optimized flows also would drastically reduce the number of types of experience that would go unused in officers' current jobs (shown on the left), roughly doubling the utilization of prior experience (shown in the dark bars on the right). Largely similar results emerged when we sought breadth instead of depth or considered possible future changes in some jobs' needs for experience or changes in the number and mix of future jobs.

As noted, RAND is conducting similar research for the rated and intelligence forces. Anticipating practicality and improved efficiency for subsequent adaptations, we believe that *development teams have the ideal scope and membership to guide such steps for their career fields,* assisted by Air Force manpower and personnel analysts. It will be straightforward to synchronize these steps with the skill pairings targeted for future senior leaders. Improvements in existing manpower and personnel data systems could facilitate such systematic development planning and then management. For example, the Military Personnel Data System has data fields (unused to date) that could hold information about jobs' demands for prior experience, education, or training and about officers' corresponding backgrounds, but computer programming and some conceptual design work are needed to enable their use. With these improvements and a model like the one described here, development teams and Air Force analysts should be able to establish well-justified targets for their individual career fields and guide the development of officers with the right match of skills for their jobs.

Next Steps for Force Development

The Air Force expects to advance force development in the coming years, greatly improving the development and utilization of highly qualified officers for senior positions and within each career field. The approaches outlined in this monograph can help the Air Force achieve these goals, but further steps remain, especially in the three areas described below.

Evaluating and Measuring Progress

Evaluation mechanisms and measures should be developed and applied to track force development's progress and effectiveness, addressing both skill pairing and development within a career field. At first, it will be useful to track

- development teams' progress in issuing developmental vectors, consistent with identified goals (especially skill-pairing goals for each specialty), and progress in establishing objectives for development within a career field
- assignment teams' success in directing officers into jobs consistent with their developmental vectors

- cohorts' evolution toward meeting career field–wide goals.

Additionally, the evaluation of selection and promotion processes is especially important for force development's success. Even if the development teams issue developmental vectors in proper mixes and the assignment process eventually places people in corresponding positions, the system will fail if people who are deliberately developed for leadership jobs and for promotion are not later selected for such advancement. Consequently, *we recommend assessment measures that track the mixes of people who are promoted to each grade, serve in command jobs, attend developmental education courses in residence, and hold other key assignments.* Do they come from the appropriate career fields? Are their paired skills consistent with established targets? Does a cohort's mix of backgrounds come closer to targets as it progresses? If too few people who are deliberately developed for advancement are selected, it means either that the wrong people have been developed or that the selection process has not valued their development adequately.

The Air Force also should see better matches between jobs' needs and incumbents' prior backgrounds. It seems practical to measure and monitor the match for the relatively smaller forces in the higher grades and the fewer paired skills at first, expanding later to monitor lower grades and consider other types of background—e.g., experience in mission operations, functional areas, organizations, and command. To enable the latter, more-extensive assessments, it probably will be cost-effective to tap the Military Personnel Data System's latent capacity. *Consistent mechanisms are needed for*

- recording jobs' needs for prior experience, education, and training[16]

[16] Personnel requisitions, submitted online and maintained at AFPC, often contain such information, but they are neither consistent nor are they presented in a manner that allows broad summaries (e.g., how many jobs require a specific element of experience, education, or training?); they do not help identify good candidates (e.g., via comparing jobs' needs with members' backgrounds); nor do they support performance assessments (e.g., how well assignees' prior qualifications match jobs' needs, overall).

- tracking individual members' accumulating portfolios of experience, education, and training
- helping commanders and mentors recommend assignments and assignment teams make assignments whose demands members meet, that are consistent with targeted developmental patterns, and that match members' preferences, insofar as possible.

Concrete improvements are needed in all three areas.

Improving Force Planning and Management

Force development proceeds within the context of force management, where broader numbers are key: How large should each specialty be? How many people are needed at each grade or skill level? What is the appropriate mix of active, reserve, and guard personnel? What is the appropriate mix of military and civilian personnel? What retention, promotion, training, cross-training, and separation programs are needed to maintain appropriate strengths?

RAND has proposed ways of improving force planning and management over the years, addressing both within-component[17] and cross-component[18] planning aspects. The FMDC could offer new and important oversight of such management, perhaps via rejuvenating the Air Force's Total Force Career Field Review, which deliberately concentrated on higher levels of force management. By component, type, grade, and skill level, it compared each career field's assigned personnel with required and authorized manpower and revealed differences across career fields. With such information, the FMDC could prioritize some specialties over others for recruiting, training, cross-training, retention, or contracting resources, for example.

[17] See, for example, Galway et al. (2005); Schiefer et al. (2006); Moore (1981); and Gotz and McCall (1984).

[18] See, for example, Robbert, Williams, and Cook (1999); Moore et al. (1996); Palmer and Rydell (1991); Gotz et al. (1990); and Rostker et al. (1992).

Palmer, Adele R., and C. Peter Rydell, *Developing Cost-Effectiveness Guidelines for Managing Personnel Resources in a Total Force Context: Executive Summary,* Santa Monica, Calif.: RAND Corporation, R-4005/2-FMP, 1991.

Robbert, Albert A., Steve Drezner, John E. Boon, Jr., Lawrence M. Hanser, S. Craig Moore, Lynn Scott, Herbert J. Shukiar, *Integrated Planning for the Air Force Senior Leader Workforce: Background and Methods,* Santa Monica, Calif.: RAND Corporation, TR-175-AF, 2004.

Robbert, Albert A., William A. Williams, and Cynthia R. Cook, *Principles for Determining the Air Force Active/Reserve Mix,* Santa Monica, Calif.: RAND Corporation, MR-1091-AF, 1999.

Rostker, Bernard, National Defense Research Institute, Charles Robert Roll, Jr., Marney Peet, Marygail K. Brauner, Harry J. Thie, Roger Allen Brown, Glenn A. Gotz, Steve Drezner, Bruce W. Don, Ken Watman, Michael G. Shanley, Fred L. Frostic, Colin O. Halvorson, Norman T. O'Meara, Jeanne M. Jarvaise, Robert Howe, David A. Shlapak, William Schwabe, Adele R. Palmer, James H. Bigelow, Joseph G. Bolten, Deena Dizengoff, Jennifer H. Kawata, Hugh G. Massey, Robert Petruschell, S. Craig Moore, Thomas F. Lippiatt, Ronald E. Sortor, J. Michael Polich, David W. Grissmer, Sheila Nataraj Kirby, Richard Buddin, *Assessing the Structure and Mix of Future Active and Reserve Forces: Final Report to the Secretary of Defense,* Santa Monica, Calif.: RAND Corporation, MR-140-1-OSD, 1992.

Schiefer, Michael, Albert A. Robbert, Lionel A. Galway, Richard E. Stanton, and Christine San, *Air Force Enlisted Force Management: System Interactions and Synchronization Strategies,* Santa Monica, Calif.: RAND Corporation, MG-540-AF, 2006.

Senior Leader Kickoff meeting regarding "Broadening the Officer Force," Air Force Personnel Center, Randolph AFB, Tx., July 27, 1999.

Space Professional Development, Web site. As of September 1, 2006: http://www.peterson.af.mil/spacepro

Thie, Harry J. Margaret C. Harrell, Roland J. Yardley, Marian Oshiro, Holly Ann Potter, Peter Schirmer, and Nelson Lim, *Framing a Strategic Approach for Joint Officer Management,* Santa Monica, Calif.: RAND Corporation, MG-306-OSD, 2005.

U.S. Air Force, *Officer Professional Development, Vol. 1, Total Force Development (Active Duty Officer),* AFI 36-2640, January 23, 2004.

Vernez, Georges, S. Craig Moore, Steven Martino, and Jeffrey Yuen, *Improving the Development and Utilization of Air Force Space and Missile Officers,* Santa Monica, Calif.: RAND Corporation, MG-382-AF, 2006.

The Women on the Porch

Southern Classics Series

M. E. Bradford, Editor

Southern Classics Series

M. E. Bradford, Series Editor

Donald Davidson — The Tennessee, Volume I
Donald Davidson — The Tennessee, Volume II
Caroline Gordon — Green Centuries
Caroline Gordon — None Shall Look Back
Caroline Gordon — Penhally
Caroline Gordon — The Women on the Porch
Johnson Jones Hooper — Adventures of Captain Simon Suggs
Madison Jones — The Innocent
Augustus Baldwin Longstreet — Georgia Scenes
Andrew Nelson Lytle — Bedford Forrest and His Critter Company
Andrew Nelson Lytle — A Wake for the Living
Thomas Nelson Page — In Ole Virginia
William Pratt, Editor — The Fugitive Poets
Elizabeth Madox Roberts — The Great Meadow
Elizabeth Madox Roberts — The Time of Man
Allen Tate — Stonewall Jackson
Robert Penn Warren — John Brown: The Making of a Martyr
Robert Penn Warren — Night Rider
Owen Wister — Lady Baltimore
Stark Young — So Red the Rose

The Women on the Porch

CAROLINE GORDON

with a preface by Louise Cowan

Ivan R. Dee, Publisher
CHICAGO

A J. S. SANDERS BOOK

To Dr. and Mrs. Max Wolf

Οὐ γάρ δὴ οφθαλμοῖσί γε ἰδόντι τούτων τῶν εἰρημένων οὐδενὶ οὐδὲν ἔστιν εἰδέναι.

Ἱπποκράτης ΠΕΡΙ ΤΕΧΝΗΣ

Published by arrangement with
Farrar, Straus & Giroux, Inc.

Library of Congress Catalog Card Number:
92-062892

ISBN: 1-879941-20-1

For information, address: Ivan R. Dee, Publisher,
1332 North Halsted Street, Chicago 60622.
Manufactured in the United States of America and
printed on acid-free paper.

Preface

"My next novel is to be a study of Merry Mont," Caroline Gordon wrote Sally Wood in 1937, referring to her family place in Kentucky. "I am going to call it 'The Women on the Porch.' I hope the title sounds sinister. It's meant to be." Then later, to another friend, she wrote again of the novel, "I'm calling it 'The Women on the Porch.' The porch is a sort of stoa to Hades. If the story has any form it is that of a myth, Eurydice and Orpheus." She did not immediately begin the novel she had anticipated, however, but instead turned to her monumental work *Green Centuries*, which, completed in 1941, recounts the epic venture of the settling of Tennessee during Revolutionary times. But she used the "women on the porch" image in a short story. "The Brilliant Leaves" (1937), depicting the inevitable masculine betrayal as leading women either to the porch or the precipice. By the time she began work in 1942 on the Merry Mont project she had described in her letters, she had already published five novels, all of them exploring in different degrees the tragic implications of the heroic calling. In this sixth novel, however, she goes beyond tragedy, deserting the daylight—and essentially masculine—realm of her ancestral heritage to enter into the nocturnal, feminine regions of the underworld, exploring a death-experience into which the psyche can enter and from which, with help, it can emerge. "I was haunted by Gluck's opera [*Orfeo ed Euridice*], both by the music and by his version of the Orpheus story," she wrote: "I was conscious of parallels between the form of the opera and that of my novel." In *The Women on the Porch*, which she completed in 1943, the fate of woman is perceived to be not necessarily either the porch or the precipice; as in Gluck's opera, the god Amor can restore to life the soul willing to take the risk of loving. *The Women on the Porch* thus marks a turn in Gordon's work: from this point on events that have been seeming to flow in one direction are shown to be headed on a quite different course.

Readers have noted Gordon's changed style and method in this sixth novel: she moves from a former naturalist technique to the oneiric and the visionary, from "linear narration" to what Radcliffe Squires speaks of as "orbicular scenes folded within scenes," or—as she herself described the new technique—interspersed sections of past and present "like 'broken colours,' to use a painter's term." In a letter to a friend she speaks of being as excited over this technical innovation as Uccello was over the discovery of perspective. An even more basic change, however, is her shift to what we might call, along with Bakhtin, a polyphonic technique. She uses the intertwined voices of her multiple interior monologues—at least eleven separate consciousnesses, in addition to the chief narrator's—as probes into a fictional universe constituted less by a central set of characters than an entire segment of human experience. This is perhaps only to say that she is writing with her eye more on the unifying inner form of the novel than on such elements as the delineation of character or sequential plot as ends in themselves.

A remark twenty years later in her "Letters to a Monk" sheds light on this quest for unifying form: "When I write a book I mull over things till what Aristotle calls 'the action' is clear to me . . . the difficult thing for a writer is to discern this action and then to make sure that every element in the novel contributes to this underlying movement." She goes on to declare that the work "is all done below ground," or perhaps is an act of what Proust calls "translation." "The work already exists in another country," Gordon says, "like a great statue which you bring over, piece by piece and, with luck, re-assemble on your own terrain." It is this growing awareness of another reality coexisting with our own and the increasing discovery of the necessarily fragmentary nature of the amassing process, the translation, that takes her writing even farther into territory not ordinarily attempted in the novel.

Responding to critics who say *The Women on the Porch* should have depicted more of the intimate life together of its main characters—Catherine and Jim Chapman—Gordon confided to Sally Wood, "None of my books ever seems round enough. They are always too lean somewhere." This "leanness" stems precisely from her concern for a greater scope and less subjectivity than the novel ordinarily provides. Her aim is hardly character portrayal. The myriad interior voices of *The Women on the Porch* make up a *heteroglossia* which is concerned to reveal not simply the psychological lives of individual personages but their placement in a cosmos of some ontological significance, in which a particular movement of soul takes place. What is emphasized here, then, is the many-voiced chorus of the feminine psyche, which, in evoking generations long dead, establishes a matriarchal order defying time, in which masculine consciousness is an intrusion and a challenge. The cosmos so construed is nonetheless epic, even if it is depicted in only one stage of epic action—the *catabasis*, or descent to the underworld. In this subterranean realm of *memoria* in which the past lives on in its ghostly apparitions to trouble the present, the central subject of this novel is still, as in all of Caroline Gordon's fiction, the heroic life and its intuitions of a future.

In *How to Read a Novel* (1957), Gordon points out that the proper hero of a fiction is a man of action who engages himself with the objective reality of an enclosing world, a world that, she maintains, "from time immemorial has been personified in the feminine consciousness." But the feminine consciousness, too, can go astray. *The Women on the Porch* is about the obsessive structures that feminine sensibility can build, into which women can escape and by which they are then entrapped. What it shows is that for the inner life to resume its flow, for the underground spring to be accessible to the psyche, for the muse once more to descend and the hero to return to his austere path, the masculine and feminine polarities within the psyche must be brought together in a unity of being. Gordon sets up typological reverberations within her novel by means of parallels and allusions to various "catabatic" situations, the most familiar of which are the expeditions to the underworld depicted in *Odyssey* XI, *Aeneid* VI, and the whole of the *Divina Commedia*, with, further, the myths of descent to Hades—Persephone's to embrace and thus placate death, Orpheus' to bring back Eurydice, and Heracles' to recover Alcestis—reinforcing the mysterious pattern finally fulfilled in the harrowing of hell. Her purpose in using the mythological and poetic references is to gather them together in a focus upon the integration of the psyche—its redemption from the dead. Hence though in this work, as in all her other novels, the large cosmos remains epic, with the hero's fidelity to his calling still the main issue, *The Women on the Porch* concentrates on a single aspect of that epic cosmos: the wedding of the masculine and feminine elements of the psyche and the journey to the underworld as requisite to that *conjunctio*. In emphasizing the marriage bond, however, Gordon is concerned not simply with domestic life or erotic fulfillment but, on one level, with the life of the soul; on another, the future of civilization.

The marriage central to the novel, however, has been violated by Jim Chapman's adultery; and his wife, Catherine, like the snakebitten Eurydice, feels the cold assaults from the regions of the dead. She seeks refuge in her remembered childhood retreat, Swan Quarter, an isolated preserve cut off from the world by a Styx-like river and guarded by a trio of women. They sit on the porch in the forbidding guise of priestesses at the entrance to an underworld: Catherine's grandmother—old Miss Kit—who, lacking the courage to marry the man she loved, dwells now in a spectral world of "presences"; Aunt Willy Lewis, a spinster who has accepted her stoic destiny of self-abnegation; and a more fragile creature of blight and shame, Cousin Daphne Passavant, frozen, twenty years ago, at the desertion of her bridegroom on their wedding night. The joylessness of these three is reflected in the black servant, Aunt Maria, whose son Jesse is serving a life term in prison, a lot Maria endures with malediction and bitterness. Catherine later ponders Maria's unnerving detachment, inadvertently characterizing her own situation: "If your heart were broken, if a great fissure came in the center of your being, you might turn your vision inward, might from then on contemplate only what could be seen in those shadowy depths. People in the outer world would become ghostly. . . ." Swan Quarter is a matriarchal realm where regret and chagrin prevail, where standing on the verge of a fearful choice, Catherine is postulant to a blighted sisterhood.

She has fled to this sanctuary, a threshold leading to the domain of shades, from a rootless society of soullessness and repetition, rushing on her flight as one plunges toward oblivion. With the black dog, Heros, at her side on the drive down from New York to this borderland between Tennessee and Kentucky—and between life and death—she feels the precious vital fluid leaving her body like ichor from her veins, sees only the "insubstantial shapes falling away on each side . . ." and thinks with Eliot and Dante, "who would have thought death had undone so many?" Dust, moths, darkness, thirst, loneliness—these mark her flight toward the dark waters of Lethe. Arriving at Swan Quarter, her way blocked by elderberry bushes with their black fruit, she abandons her car and strikes out on foot, with the dog, snakelike, cutting through the underbrush. Plunging into the thicket, her legs bloodied by thorny vines,

> She ran forward, falling to her knees once when she stepped on a loose brick. She picked herself up and saw at the end of the green tunnel the gray, spreading bulk of the house. Women were sitting on the porch. One was old and stout and wore a lace cap on her white hair. Another woman, thin almost to emaciation and with black, restless eyes in a sand-colored face, sat close beside her, book in hand. On the steps below the two a wiry, middle-aged woman seemed just to have dropped down to rest.
>
> The sun was sinking behind the trees. The faces, the immobile bodies swam in western light. For a moment it seemed to her that she had never seen these women before. . . .

Despite their seeming passivity, these women are, in a sense, the ancient Erinyes, guarding their stoa to the underworld and, as at Colonus, the doorsill to their ambiguously sacred grove, which can prove either curse or blessing. These "great ladies," as Sophocles speaks of them, represent aspects of the feminine psyche that, when injured, become furies; as the archetypal psychologist James Hillman points out in *The Dream and the Underworld*, the Erinyes possess a chthonic aspect, which "refers in origin to the cold, dead depths and has nothing to do with fertility." In the porch of this ancestral home Gordon depicts the vestibule to hell; Dante's trimmers are here, those who have not dared exist. And the women in this territory—virgin, "unravished bride," widow, mother—have made the great rejection. All are captured in a state of stasis, imprisoned in their own furious refusal to yield to the deepening of the soul that suffering and loss offer as recompense.

Symbol of this preoccupation with non-existence is Daphne's mushroom collecting: in gathering to herself the *amanita phalloides* and the *trompete du morte* she has overmastered both love and death. Hillman reminds us that mushrooms, for forest people, "are the souls of the dead springing up into the land of the living" and that in Italian folklore, the spot where a mushroom grows indicates the "planetary figures or archetypal bodies of the underworld." Daphne is a kind of queen of the underworld, a Persephone gathering her dead souls as though they were flowers and, alone in her room at night, telling them over and fingering their strange substance. She struggles to find a word to describe a specimen she holds in her hand: its color is "neither gray, nor green, nor yellow, but some unnamed color, a color a plant might develop growing in an underground cavern, or in another world."

On one level, certainly, *The Women on the Porch* is a study of the dimensions of the feminine—not only the living, but also the "presences" at Swan Quarter, seen by the grandmother and felt by the others. Aunt Willy has come to believe that those people "could get into any house—*if there was somebody to let them in*." Catherine remembers the shades that have always been at the house; she has known them from childhood. They had been then "only companions whom one could not conveniently address." But after she was a grown woman, "they had seemed at times to menace or at least to prophesy evil." And now, as she thinks, they drive her from the house to meet, in the woods, her lover Tom

Manigault. Old Miss Kit intuitively knows that Catherine is on the path that leads to the spring. "But there is some one there before her, the woman who is always on that path," the place where she, Miss Kit, had stood so long ago waiting for the man she loved and whom she lacked the courage to marry. "They have been here all the time," she now knows. "I cannot think why I never saw them before. They have been with me all along!" The veiled woman whose face she cannot see is both herself and death; and yet, as Hillman comments, the persons in dreams (and visions) are not mere images of oneself or of the persons whom they seem to represent: "They are shadow images that fill archetypal roles, they are personae, masks, in the hollow of which is a numen." These numinous presences in Gordon's novel, pervading the liminal situation that the underworld represents, have the power to lead to the "lady tree"—the tree by the spring—all those who are willing to come.

In her state of blank despair, Catherine is at once Eurydice, Alcestis, Psyche, and Persephone, separated from her husband, dying for him, and continuing to search for him, though summoned to wed Hades—making her journey to the underworld both to rescue and to be rescued. She has been seized by what Hillman calls "the Persephone experience," when the soul feels itself "caught in hatefulness, cold, numbed, and drawn downward out of life by a force we cannot see . . . We feel invaded from below, assaulted, and we think of death." Catherine's kinship with Persephone is indicated by a silver compact her husband had given her, embossed with sheaves of wheat and round fruit (like the pomegranate), which falls out of her pocket on her first tryst with Tom but which she later recovers. But there remains the Psyche urge, the desire in spite of hurt to find her lost love, and the Alcestis impulsion to save her husband: in a "grisly" dream she and another woman are descending with a man into a long, dark tunnel:

> the man was dead, or had been dead and now, called back from the grave, hovered between life and death. Ahead of them in the vast, shadowy tunnel other people were busy with certain operations. When they had finished those operations, the man

> would be consigned to another grave, from which, it was hoped, he would rise. But in the meantime he walked beside her and kept his frail, intolerable hand on her shoulder. She was about to shake it off when somebody on ahead called back to her that she must be vigilant, that the man's safety depended on her alone.

Catherine has thought herself free to leave an unfaithful husband; but the world now available to her is either one of non-life, as manifested by the women on the porch, or of pseudo-life, as demonstrated in her desperate sexual encounters with Tom Manigault. She tries to persuade herself that to cast her lot with Tom would be to dwell in these lower regions in a natural bliss as in an Elysian Field; she could, she thinks, embrace the life of nature for which she feels herself intended. But there is to be no such earthly paradise for Catherine. Tom's manhood has been wounded from his childhood by his mother's lack of love; and Catherine comes to see that no woman will ever be as attractive to him as is his hatred of his mother.

The nadir of this non-existence, the lowest circle of this underworld, is depicted in the sacred image venerated by Tom's mother, Elsie Manigault. Forced by the war in Europe to return to the States and to stay in Big Pond, the plantation adjacent to Swan Quarter, Elsie is bored and volitionless, since, contrary to what her son believes, she is incapable of either love or hate. No longer on the porch but caught in the frigid bonds of narcissism, she turns from diversion to diversion, finally seeking sustenance in the memory of her dead father, whose portrait sits in a heavy silver frame on her desk. But the reader knows that he was a dishonest and avaricious man, with neither talent nor competence: he had only "the wisdom of an old tired organism,"

> of the old possum that lies all night in its hole rather than drag its crippled leg over the snow to the henhouse, of the hawk that, discerning with fading eyes the flash of wings through the leaves, yet clings to its perch on the dead pine, preferring the faint pain of hunger which it has had for a long time and will have until it drops shrivelled from the bough, to the agony of

> the moment when, having struck and missed, it must beat upward through empty blue.

This is a controlling image in the novel, the opposite pole from the heroic—the cold Dis of this underworld. It represents the destiny of all those who, in a refusal to risk, make the decision to stay with what they have, in a state of immobility, even at the cost of life. The image, as a possibility, is as applicable to Jim Chapman and his New York group of friends—the intellectual cohorts who have rejected the muse and made of the city a wasteland—as it is to the women at Swan Quarter who are trapped in a vestibule of the underworld. For Chapman, too, even though he teaches Dante in his New York university, has wandered into a dark wood where ego has replaced psyche, where death has replaced life, and the muse is silent.

If the portrait of Elsie's father exemplifies the lowest realm of this cosmos, then one might say that its highest region, though manifested neither in the wasteland of New York nor the underworld of Tennessee, is indicated in the lines of a poem twice quoted by Jim Chapman, the first time to a momentarily befriended homeless man and the second to a former poet and colleague. The poem is the *Pervigilium Veneris* (*The Vigil of Venus*), a late Latin poem lamenting the silence of the muse. This lyric sequence, in its first stanzas an expression of the *carpe diem* theme, deepens into a general lament for the loss of love and song. The poet speaks in his own voice, out of his own silence, comparing himself to the Amyclae, the citizens of a village destroyed because they were silent, in contrast to the violated Philomela, who, metamorphosed into a swallow, makes of her injury and grief an utterance of beauty. The context for the lines quoted by Chapman is the last stanza of the poem, which Allen Tate, Gordon's husband, translates as follows:

> She sings, we are silent. When will my spring come?
> Shall I find my voice when I shall be as the swallow?
> Silence destroyed the Amyclae: they were dumb.
> Silent, I lost the muse. Return, Apollo!
> Tomorrow, let loveless, let lover tomorrow make love.

In a note to his translation, included in his *Collected Poems*, Tate makes an instructive comment on the passage: "this long, gentle meditation on the sources of all life [the generative love of Venus genetrix] comes to a climax in the poet's sudden consciousness of his own feeble powers. When shall I, he says, like Philomela the swallow, suffer violence and be moved to sing?" Tate saw in this lyric poem the cry not only of the poet but of the people who lose their voice; he must then ask, "is the poem not telling us that the loss of symbolic language may mean the extinction of our humanity?" In *The Women on the Porch*, Jim's twice quoting from the poem indicates that not only is the marriage of Catherine and Jim at stake, but, in the general silence of the muse, the life of culture.

This ultimate reach of the heroic quest, the high calling of the poet, which is an overt theme in Gordon's later novels, is here used in her work for the first time. The poet's task, she maintains, is to keep his eyes on the heroic life, something Chapman in his disillusion and cynicism has failed to do. He is a failed poet; and his intellectual discipline, history, has become for him a source of despair, the emptiness of his life leading to his extramarital affair. He has cut himself off from the springs of both love and poetry. The loss of Catherine shocks him out of his routine; and in a state of drunken despair he begins his search for his true identity. According to Jung, Hillman's precursor, this painful quest for the self must lead to a confrontation with the "shadow," its unacknowledged counterpart if the self is to be integrated. Caliban for Prospero is such a dark twin, and Poor Tom assumes this role for King Lear. Catherine Chapman encounters her shadow in the rejected and unlovable Cousin Daphne and shrinks from any comparison to her. Similarly, it is in the repulsive "hobo," a streetperson, formerly a scholar, that Jim must find his repudiated self, the disgusting and pathetic mortality which has gone unacknowledged in his life. The "bum," as Jim refers to him, is a "thing of darkness" that must be accepted as his own by a purely gratuitous act of kindness before the way will moderate its downward plunge. But once Jim withdraws from the apartment of the woman of his attention, not so much with resolution as with certainty that he will not again enter "that thicket," the chaotic events of his life begin to trace out an unexpected design. With a minimum of volition, he boards a train bound for Carthage and Swan Quarter. It is a journey of transition; in a surrealistic vision of the future, he sees the boy across from him stand on the seat, step into an airplane, and sail into the sun, disembowelling himself and scattering from the gaping wound the fruits of technology to fall upon the infertile earth.

Chapman's scorn and hatred of technology—his disguised fear of the future—are repeated later at Swan Quarter in his admonitions to one of the presences, a spectral frontiersman whom he encounters after an event of violence sufficient to shake him from his torpor. The recognition takes place when, exhausted and defeated, he makes his way to the still black pond in the woods, turning instead toward the spring at the "lady tree," seeing on the other side of the pond a shadowy figure in a buckskin coat stopping for a moment to rest. Gordon's technical artistry is made manifest in this remarkable scene, for which she has prepared the reader early in the novel in Aunt Willy's memory of recounting to Jim the story of Old Irish John Lewis and his son, her great-great-grandfather, the first settler of the land. Jim's last attempt to retain his daylight self, his carefully preserved ego, occurs in his confrontation with this phantom who has emerged from the abyss and whom Jim recognizes as the son of Irish John. In an act of hybris, an attempt to change history, he counsels the shade against colonizing in this territory: "I would advise you not to settle on this land," he says. "The land is cursed. It is an old land, ruled by a goddess whose limbs were weary with turning before ever Ireland rose from the sea. An ancient goddess whom men have wakened from an evil dream." The man impassively cooks and eats his meal, rises, and, despite Jim's increasingly frantic warnings about the technological demons that lie ahead in his future, goes impassively on his way. His bold gaze leaves Jim stricken with a sense of his own impotence and falsity. This encounter is Jim's visitation by the presences, and in it they correct his view of history: history is made by the unheeding heroes who have no care for the ultimate outcome of their struggles, who are impelled by a divine call, like Rion Outlaw in

Gordon's preceding novel *Green Centuries*, to make their westward journey. The resolution of the novel is swift and complete: both Catherine and Jim have been purged of their false selves. Orpheus has found Eurydice and discovered her to be Alcestis; she has guided him, as her dream had warned her she must, through the tunnel. In an infusion of grace, Chapman abases himself, kisses the instep of Catherine's foot, and prepares to flee the underworld with her "when it's light." They are now looking toward morning, then, whereas our first view of Swan Quarter was at dusk.

This ending has puzzled readers, even such astute ones as Andrew Lytle. The reconciliation between husband and wife comes, on the natural level, out of the blue. But in the strictly restrained events of the last few pages Gordon traces the workings of what to Sally Wood she called the "circumnatural," describing this quality as a sense of "the intangible verities that lie about us and are yet not supernatural." She is no doubt speaking of the dimension of myth and archetype, the ancient psychic patterns that still prove valid in any experiential search for human wholeness. And though her eventual *telos* is the grace-filled canvas of her later novels *The Strange Children* (1952) and *The Malefactors* (1956), which increasingly make room for the overtly supernatural, nevertheless, the mythic level of *The Women on the Porch*—specifically, the soul's defining itself in terms of darkness and death—remains a necessary stage in the course of her completed *oeuvre*. But *The Women on the Porch* is important not only in the author's body of writing and not only in the larger canon of Southern literature. It is an important milestone in the course of the novel, for it enlarges the scope of that medium so that, finally, its limitations of facticity and rationalism are set aside, to allow expression in our own day of the mysterious borders of life dealt with perennially by the great literature of the past.

Dallas, Texas LOUISE COWAN

The Women on the Porch

1

THE SUGAR TREE's round shadow was moving past the store. At five o'clock when the first leaves were withering on the burning macadam the storekeeper raised his eyes to the fields across the road. The heat rose somewhere between the road and those distant woods. Always at this hour he looked, expecting to see it rise out of that far cornfield and always when he looked it was there. Only a light shimmer now above the green, but the shimmer would deepen as the field brimmed over. In a few minutes the first waves would beat against the porch. He got up and, walking to the end of the porch, lifted the lid of the red metal ice-chest.

"How about you, Ed?" he asked.

The man at the other end of the porch leaned forward, felt in his trousers pocket until he found a nickel, and pitched it to the storekeeper.

The storekeeper brought two bottles up out of the chest, removed their caps and handing one to his friend sat down again. He raised his bottle to his lips. The liquid was no cooler than well water. He spat it out into the dust. "Truck ought to be getting here," he said. "I ain't got ice to last longer'n six o'clock."

"There's somebody at the top of the hill now," the other man told him.

They listened. The storekeeper shook his head. "It ain't a truck."

The legs of Ed's chair hit the floor. He bent forward, his eyes fastened on the ribbon of road where it showed under the dark trees.

"Whoever it is is a-*comin'* now," he said.

Far away the sun struck a blinding gleam of light out of metal. The car sped behind some trees and came in sight again: a green convertible roadster.

It sped down the hill, flashing in the sun, seemed about to pass the store and stopped short.

The two men stared at the tracks the tires had made in the dust. "You got good brakes," the storekeeper murmured, and rose to his feet and walked to the edge of the porch.

"Evening, ma'am," he said.

The woman in the car turned her head. Her lips parted stiffly. "Fill her up, please," she said in a voice as dry as the dust that lay thick on her face, her hands, her long, light brown hair.

The storekeeper inserted the nozzle of the hose into the gas tank and walked around the side of the car. "You come far?" he asked.

Her eyes went to the bottle he still held in his hand. "Will you give me a Coca Cola?" she asked.

Ed was up off his chair and lifting the lid of the chest. "Truck's late. It ain't as cold as it ought to be," he said as he extended the bottle.

Her lips, grimacing, made creases in the dusty skin at each side of her mouth. She tilted the bottle, gulped half its contents, thrust it back at him and, leaning forward, started the motor.

The storekeeper jumped to one side as the fender cleared him and, shading his eyes with his hand, looked after the retreating car. "I know that woman," he said.

Ed laughed. "That's a New York license. Take her out last time you was in New York?"

The storekeeper shook his head. "She's a Lewis. I can tell by them yellow eyes."

Ed's gaze shifted from the bridge to the woods on the other side of the river. "I ain't been over to the Lewis place in ten or fifteen years. The old lady still living?"

"She'll live to be a hundred," the storekeeper told him. "Her daughter runs the place. Miss Willy. But I ain't seen her in a long time." He slapped his haunch. "Ed," he said, "that girl in that car is Agnes Lewis' daughter."

"There was three of them young Lewises," Ed said, "Agnes and Jack and Willy. Jack broke his neck, riding a wild horse. I was sitting here on the porch the day a nigger boy come with the news."

They walked back to the porch and sat down. The storekeeper sat with his hands on his knees, his eyes bent on the dusty floor, as if he read there the Lewis genealogy.

"Agnes . . ." he murmured. "Married a newspaper man and went to live in New York City. Had a baby that she used to bring down here summers. Named Catherine, after the old lady. The girl in that car, Ed . . . But she ain't been here in many a summer. Wonder what brings her now?" His eyes sought the woods on the other side of the river. "She'll back that car out once she gets in there. It's lonesome, the old Lewis place."

The woman, settling back against the leather cushions, heard her mother's surname called and then the men's voices were lost in the rattle as the wheels struck the loose planks of the bridge. She raised her body up a little and looked down over the hand-rail. The water along here was always colored green by moss. Farther down, where it ran over pebbles, it was clear. "As clear as your eyes," Mammy used to say. That was when she was little and they would bring her down to the creek to wade . . . Another rattle and the car was off the bridge. Before her the road rose steeply. Pigeon Hill. If you kept on over the hill you would come to Carthage. Carthage where everybody who walked the streets knew you. But you would not stay in Carthage. You would drive on. To Nashville. You would drive out of Nashville on the Murfreesboro pike and then—she had done it once in half a day—Atlanta. And after Atlanta little towns where dogs slept in the dust around the courthouse square and then Mobile and live oaks and the road straight to New Orleans. You would drive through the swamps and over the Honeysuckle Bridge, strike Gentilly and turn down to the river. There would be negroes loading cotton at the docks. You would drive up under the shed, past the negroes, past white men who might put out hands, then, before any hands could clutch its sleek, gliding sides, the car would plunge down, past rotting green piles and the yellow water would swirl over your head. But it would not stay yellow. Swirling over your head, it would change, to blue, to green, to purple laced with foam. *Where'er thy bones are hurled.* People buried in the sea have shifting coffins. You cannot ever lay flowers on their graves.

That is not the shortest way to the sea. You lack geography, Catherine.

"I have to go the way I know," she whispered and aloud: "Swan Quarter. We're going to Swan Quarter."

She swung the car off the macadam on to a side road, so suddenly that the black dog, lying beside her, was catapulted to the floor. He whimpered and, climbing up on the cushion, reared himself against her and licked her arm.

She put her hand on his head. "Be good," she said still in a whisper. "Be good now."

That first night, in Staunton, Virginia, the clerk had not wanted to let the dog into the hotel. It was three o'clock when she came in, holding him in her arms. She had intended to drive all night until, crossing some street car tracks, she had felt the wheels slip sidewise and at the same moment had seen at her right a dark, stumbling form. She had driven on swiftly but at the next corner had stopped, had taken her hands from the wheel and had sat there a few minutes, her head resting on her folded arms. It was the first time she had stopped since she left New York, except for gas and the Ladies' Room and once when a motorcycle cop drew up beside her, talked to her for a few minutes and let her go.

She had left New York at five o'clock in the afternoon. It was because of the letter her husband had written another woman. When she came into the apartment, a little before five, it was there. She had crossed to the closet to hang up her jacket when she saw it, fallen on the rug just outside the door. She stooped to pick it up. She saw *Dearest* and the words she would not now allow to echo in her memory and her hand drew back and she straightened up.

It was then that she had become aware of the hush. Out on the street and in the kitchen where Lily was getting dinner, sounds went on, but in that room there had been only silence. That was what she remembered of that time, the stillness and then bending again to pick up the letter and after that walking very quietly over to the desk to lay it down and standing there a second while her eyes, which felt hot and bright and could move without making any noise, flickered about the room.

A door opened. Lily called out that she was going to run down to the delicatessen to get some croutons for soup. It was only then that she remembered that the Wares were coming to dinner. When the door had closed behind Lily she went to the telephone and called a taxi and, moving very quietly so as not to break the stillness, began taking clothes down from the racks. She had two bags packed by the time the driver came. It was just after she got the two bags into the hall and was waiting for the driver that Heros whimpered from his basket and she knew that she would have to take him with her.

It was raining when the cab drew up at the garage. The small rain, falling so quietly, seemed ichor from her own veins. That feeling of strength ebbing with some precious fluid out of the body had been so strong that she could not find words to thank the boy who brought her car out, only mouthed at him as she slid behind the wheel.

After she had passed through the Holland Tunnel and was on the highway the rain had stopped falling. There were many cars, people starting out of town for the week-end, but it was as if there were nobody on the road except herself. The dark drew in. The landscape receded. She heard the throbbing of the motor and then she could not hear anything else, as if she and the car were one engine, moving steadily on through the night. It had been like that all the time, not only on the dark highway but in the cities: Trenton, Philadelphia, Baltimore, even Washington where traffic whirled out like the spokes on a wheel. She would drive in and the people and the lights and the other cars were bright, insubstantial shapes falling away on each side of the engine that must go throbbing on. But after Washington trees closed in, the road grew narrower, and then it was mounting, with always a curve ahead, until here in this strange town the dim shape had stumbled away just in time and she stopped the motor and put her hands down on her folded arms. She sat there, her forehead pressed against the cool wood, her fingertips holding her eyelids shut while the throbbing died out. And then she raised her head, feeling sick and weak now that it was gone and the street was dark around her and she knew that she could not go any farther that night.

She turned back into the main street and drove past silent houses until she came to the sign between two murky street lamps. There were steps and then a lobby, half-lit and smelling of insecticide. The clerk sat at a table under the one light, shooting craps with a negro bell-boy. He saw her and stood up, shaking his head. "Lady . . . Dogs not allowed . . . It ain't *my* idea . . ."

She stood there, holding Heros in her arms, and looked at him.

He came closer, shoving the hand that still held the dice farther down into his pocket. His blue eyes were sharp, his whole sallow face tight and then his lips parted; he pointed past her. "Doc Reilly across the street there's a vet. Maybe he'll take him."

The bell-boy had slipped out from behind the table and stood a few paces behind the clerk, watching. He said softly: "He ain't going to like being waked up this time of night."

She stepped forward, past the clerk, past the negro boy and stood in front of the desk. "A room and a bath, please."

The clerk walked slowly behind the desk, took a key down from a board and turned to the boy. "Three hundred and six . . . Get the lady's bags, Joe."

It was only as they were going up in the elevator and the smell of insecticide grew stronger that she thought that the place might be a fancy house. What was it Jim had said that time they got into the queer place late at night? Bedbugs don't *prefer* the flesh of whores . . .

The runner in the hall was dirty and ragged but the room he showed her into was clean. The door shut. She walked over and sat down on the bed that was covered with a white counterpane. She was still sitting there when the boy came again, bringing a ham sandwich and a glass of milk on a tray. "I didn't order that," she told him. "No'm, but Mister Ellis say it's so late . . ." She looked into his intelligent, dark eyes and knew that if she asked him to take Heros out on a leash he might walk him carefully around the block and return him safely. But her throat was dry; she shut the door and went back to the bed.

Heros was puzzled and crouched down in front of her, his head on his paws, whining a little. She could feel his eyes bright on her face and she made the room dark and sat there looking out of the window. The big, white street lamp pushed up into the boughs of a maple tree and had a thousand moths fluttering about it. In the light the leaves looked sleek and wet. Sometimes moths, exhausted from fluttering, would fold their wings and rest on the broad green leaves.

Heros stopped whining and got up on the bed beside her. The lamp did not go out all night but after a while it got dim and the leaves around it did not shine, though the moths still fluttered. She knew it must be near morning and lay back across the bed and shut her eyes. She woke, dreaming that she was being buried but it was only Heros, crouched on her breast for warmth. She pulled the counterpane up around both of them and slept again and was wakened by the bell-boy knocking on the door; it was the last call for breakfast, he said.

She had thought that she could not get up, that she would lie all day in that room whose peeling, yellow wallpaper had repeated upon it at intervals a hand thrust through a wreath of blue ferns. But after a while the room grew hot and then that tremor of the nerves began again until finally her whole body was shaking. She drank some bitter coffee the boy had brought, and bathed and dressed. Once on the road the shaking stopped. And after she crossed the state line at Bristol she got again the feeling that she had had the night before of being one with the engine. She could have gone on another hundred miles tonight—but that place back there was Willard's store and this was the road to Swan Quarter!

The road dipped suddenly between high banks. Elderberry bushes brushed against the sides of the car; she could have put out her hand and plucked the hanging black fruit. She slowed up for a projecting rock and stopped. A little farther on a tree was suspended between the two banks: a young tulip poplar, its leaves still fresh and green. She got out of the car and tugged at the trunk but the ball of earth that clung to its roots made it too heavy to move.

There was a rustling in the bushes on the bank. The dog's snake-like body showed through the leaves. His tongue lolled out between his black lips. He seemed to be smiling. "You look like the Cheshire cat," she whispered, and found the sound of her own low voice intolerable and caught hold of an overhanging bush and pulled herself up beside him.

Standing waist deep in buckberry bushes she looked at the car. She could back it out of the cut but if she got hung up on the bank she would be worse off than she was now. She shrugged her shoulders and struck off along the road, the dog at her heels.

It went on for a while between high banks, then flattened out. She came to a wet weather branch, crossed it and glanced back. Even here in this damp earth there were no tire tracks, no sign of any other vehicle's passing. She put a hand up and wiped moisture from her upper lip. Was this the road to Swan Quarter? That store back there? It was Willard's store, surely, set under the big tree with the turn-off just beyond the bridge and the water flowing green under the bridge . . . But it had been a long time since she had seen that store and she had come so far and had slept only those few hours . . .

Off towards the creek doves were moaning. She stood still to listen. In that instant the road disappeared. In its place was left a plantation of young dogwoods fringed deep with elderberry and sumac. She hesitated, then plunged into the thicket. The elder bushes were tall, the berries ripe to bursting. She bent her head and keeping one hand up to ward off the hanging clusters plunged on. Some vine that trailed from all the bushes kept curling about her ankles. It was barbed; she felt her stockings ripping; finally blood oozed from her torn legs.

The way grew more open. There were only the trunks of the saplings to dodge now. The vine, with no bushes to trail from, swayed on the ground. She went more carefully, lifting her feet high to avoid the loops. The dog, who had been running ahead, suddenly turned and looked back at her. He had come to a fence. She looked up and saw, beyond the dogwood leaves that were edged with light, masses of red and purple bloom.

The fence was five-barred, broken a few feet from where she stood by a long gate. The dog had already found the gate and scrambled through. She pushed it open and stood on thick, neglected grass at the end of a lawn.

Before her an enormous clematis vine reared itself seemingly unsupported. But she knew that it depended from an ancient iron trellis. And the red was the bloom of crape myrtle. She thrust the masses of clematis aside and was on a sunken brick walk between tall crape myrtle bushes.

She ran forward, falling to her knees once when she stepped on a loose brick. She picked herself up and saw at the end of the green tunnel the gray, spreading bulk of the house. Women were sitting on the porch. One was old and stout and wore a lace cap on her white hair. Another woman, thin almost to emaciation and with black, restless eyes in a sand-colored face, sat close beside her, book in hand. On the steps below the two a wiry, middle-aged woman seemed just to have dropped down to rest. Her forehead, even the fine, brown hairs of her head, glistened with sweat. Her hands, loosely clenched, swung between her spread knees.

The sun was sinking behind the trees. The faces, the immobile bodies swam in western light. For a moment it seemed to her that she had never seen these women before; and then she knew them for her aunt, her grandmother and her cousin and she called out their names and ran towards them.

II

Willy Lewis woke while it was still dark. At first she did not know where she was and then she heard the labored, irregular breathing from the other bed and realized that she was sleeping on the cot in her mother's downstairs bedroom. She had had it brought down from the attic and put in there on Saturday a week ago.

Her whole thin body suddenly went taut. As she turned over in bed and, lying face downwards, reached up and grasped the low headboard it seemed to her that all the blood in her body was concentrated in her fingertips. She had had this impulse to reach out and grasp something for a week now, whenever she remembered what had happened Saturday. She had been in the stable lot, seeing to something, and around four o'clock had come into her mother's bedroom. The old lady, just waking up from her nap, sat up in bed. There was a smile on her face. "The jasmine is blooming early this year, Alice," she said.

Old people always called you out of your name by a generation. Willy was quite used to being called by the name of this aunt who, fifty years ago, had died in her twentieth summer. She went to the washstand and poured some water into the basin and was washing her hands when the voice spoke again, more sharply.

"Have you picked the raspberries?"

Willy's hands that had been moving through the water stiffened. "No, Mama."

The old lady was out of bed and coming towards her. "They ought to have been picked while the dew was on them. You'll never get them chilled now."

Willy dried her hands on the towel and turned around. "We haven't got any raspberries, Mama . . . Would you like some peaches cut up, with cream?"

"In *May*?" her mother asked and broke into sudden, cackling laughter. "I don't see how you can be so forgetful and the house full of company."

She had taken up her white skirt from where she had thrown it on a chair and was trying to put it on. Willy went to her and, standing behind her, slipped it down over her hips and started buttoning the placket. "There isn't any company, Mama. Cousin Daphne doesn't come till next week."

"Daphne!" old Catherine said and jerked away from her. "I hope not. She'd have to sit on the shelf . . . Which room did you put Mrs. Roswell in? Cousin Tom won't like it if the baby cries all night."

Willy glanced up over her head. The room above, her own bedroom, had been Mrs. Roswell's room, she had always heard. But it had been seventy years since Annie Roswell had visited here. She had died on her last visit, of typhoid fever, and she and her baby, Sara, were buried in the graveyard on the hill.

The old lady was fumbling with her shirtwaist. "Mrs. Roswell and the two boys," she whispered. "Harry's coming Sunday . . ."

Outside in the grass there was a light rustle. Willy felt the nails dig into the palms of her clenched hands. Were they coming, those people who had been dead for years?

They *could* come, they could get into any house—*if there was somebody to let them in.*

Her mother had twisted about and was looking up at her. Her lips, red in spite of her ninety years, still curled in irritation but the eyes that fixed Willy's face were frightened. "Alice," she whispered. *"Alice?"* Her lips trembled; she was crying.

Willy bent and put her arms about her. Under the thin lawn of the shirtwaist the old shoulders were firm. A wisp of white hair, escaping from its knot, curled on the neck. Willy, bending lower, caught the mingled odors of orris root and clean, withered flesh. She tightened her clasp. "Willy's got you," she murmured, "Willy's here. *Willy* . . ."

Her mother stood in her embrace for a moment and then she was drawing away, making a little fanning motion with her hand. "It's a little warm for demonstrations of affection . . . Here, button this, will you? . . . How anybody can get into these new-fangled clothes . . ."

They had gone out on the porch and Willy had read to her out of a magazine, a story about a secretary who spent all her savings on clothes and went to a sea-side resort, determined to catch a rich husband. She fell in love with a young man but turned him down because he was poor and then when she got back to her office found that he was her new boss, an enormously wealthy young man who had gone to that resort hoping to escape the pursuit of society girls and their mothers. The old lady yawned and said they were a couple of fools and deserved to spend the rest of their lives together. She had seemed quite herself all the rest of that day, ever since, in fact.

Willy turned over and lay looking out of the window. It had rained during the night. Between the dark oak boughs the eastern sky was curdled like a bowl of milk. But the curds were growing rosy, sliding away from the sun that was already a hot, bright point of light. She ought to get up. It would be day soon. This cot, mercifully, did not squeak. If she tiptoed out of the room her mother would not wake but would sleep on until breakfast.

She sat up and, stretching her arms high over her head, yawned noiselessly, then glanced over at the other bed. The old lady was sunk deep in the feather mattress. Only the curve of her white hair and the tip of her nose showed above the pillow. In the last hour her breathing had grown easier. Mama's mind was so strong. She might have momentary lapses but she would never turn foolish the way some old people did—Cousin Molly Overton, who kept asking you who she was; Miss Bessie Froude, who at the age of eighty had taken a notion that she had a young baby. No, Mama would never be like them. It had been silly of her to get so frightened.

There was the sound of a door slamming downstairs and then the clatter of pans. Maria had come. If she hurried she might have time to feed the chickens before Maria got breakfast ready. With the morning work out of the way she could get into the lot while it was still cool, perhaps walk from there over to Shannon's. She had made up her mind to buy the alfalfa seed but she would like to know what Mr. Shannon thought of the price. Tom Manigault would not take advantage of his cousin, of course. Still, thirty cents seemed high, or perhaps she was mistaken when she thought she had heard of some seed selling for twenty-five cents last year. It was hard for a woman to hold all those things in her head. Sometimes she wished she were back in the old days when Jack ran the farm and she had nothing to do but wait on Mama.

Her lips tightened. No. If Jack were living now she would not want him to run the farm. It had been wrong of Jack—it had been *wicked*—to put the mortgage on the place, handing Mama the papers to sign without even telling her what they were and after he got things into his own hands selling off the land across the road without a by-your-leave to anybody. It was not having that pasture land that had made it so hard for her these last few years. Mr. Shannon said that she had no choice but to go into hogs. And it sounded reasonable. When you had all those rich bottoms, and even Mr. Shannon admitted that the bottoms were built up now, you *had* to raise corn and when you raised all that corn you had to put it to the hogs or else give it away for fifty cents a bushel as so many people had done last year.

It was strange, though. Mama had been almost wild when she discovered that there was a mortgage on the place but she had never really been mad with Jack, not half as mad as she had been with her, Willy, when she sold the dogwood trees for shuttles. And yet they could never have got through the last two years without that money.

She swung her legs over the side of the cot and was on her feet in one quick motion. She must get dressed and get downstairs. There was still time to tell Maria to make spoon bread for breakfast, and they might open some of that strawberry jam. Lying there, worrying about her own affairs, she had completely forgotten that her niece, Catherine, had arrived last night.

As she tiptoed out of the room and up the dark stairs she wondered what could have brought the child here, without a letter or even a postcard to tell them that she was coming. She had thought when she looked down and saw her there at the foot of the lawn, her hair all fallen down, her legs bloody from the briars, that something terrible must have happened: Jim Chapman was dead or he had deserted her or she herself had fallen into some unimaginable evil plight. And then she realized that she looked that way because of the time she had had getting through the thicket. The old lady had been sharp about that. "You see what happens when you let a whole grove of dogwoods grow up at your front door, Willy." "It's not quite at our front door, Mama," she had told her, "and you know we never use that road now. It would be foolish to keep it up when it's so much easier to go out the back way."

She let herself into her own room and, going behind the screen that stood in front of the washstand, slipped off her nightgown and poured water into a bowl. As she passed the cold, wet cloth over her body she found herself thinking of her sister, Agnes, Catherine's mother. She sometimes had to remind herself that Agnes was dead. It was as if she were living on there in New York, only they did not get letters from her as they used to do. Catherine had been good about writing the first two or three years after her mother died, and for a while after she married. Then the letters had gotten fewer, until finally they degenerated into notes accompanying Christmas and birthday presents. You could not blame the child. It was hard to keep on writing letters to two old women off in the country whom you had not seen for years.

She took up a towel and began drying herself. I *am* old, she thought. In a few years, five, anyhow, I'll have the change of life. She put the towel from her and stood, her hands akimbo on her spare hips, then drew her body upright. I'm strong. And I don't feel old. I don't feel as old as I did ten years ago. That's because Jack's gone. When he was alive there wasn't anything for me to do but dry up. Now I have to see to things. She leaned her forehead against the towel where it hung over the screen and pressed the rough, damp folds to her trembling mouth. *I wasn't ever really glad, Jack. I'd give it all up to have you back* . . . Her brother's face rose before her, the light brown eyes that people said were Lewis eyes, the wide mouth. He was leaning back from the table after a late supper, laughing. But the sturdy, athletic body did not draw away from the table. It *rose*, high into the air, a terrible arch that might have been his death convulsion, as it nearly was; the doctor said that his neck must have been broken the instant the horse fell and he struck the ground.

She wiped her face and going over to a chest of drawers took out some underwear and a blue percale house dress. She was buttoning the dress when she looked in the mirror and saw that it had a blackberry stain on the skirt. She took it off and put on instead a blue linen with red buttons down the front. There was company in the house: Catherine and Cousin Daphne. She might as well look her best, "though I can't ever look very well," she thought, leaning forward to stare at her small, deeply set eyes, her pale lips that closed over slightly protruding teeth.

As she started downstairs she heard a scratching in the hall below. The sleek, black dog sat before the closed dining room door. She stooped and patted his head. He fell on his side, cringing and thumping his tail on the floor. She opened the door. He dashed into the dining room.

Catherine was already at the table. She wore a pale colored dress of some rough-woven stuff and sat a little turned away from the table, leaning her head on her hand. Her smooth, light brown hair fell forward in a curve, hiding her face. She seemed to be gazing out of the window.

"Child, what made you get up so early?" Willy asked.

Catherine turned around. "I heard Aunt Maria come. I thought I might as well get up."

They sat down. Maria came in from the kitchen with Willy's coffee and a plate of hot cakes for Catherine. Catherine shook her head, holding out her coffee cup. "Aunt Maria, you make wonderful coffee."

Maria silently took the cup and left the room.

Catherine looked over at Willy with a faint smile. "Aunt Maria thinks I am too thin. 'Wormy' was the word."

"It's stylish to be thin, isn't it?" Willy asked.

Catherine laughed, proffering the dog a piece of battercake on a fork. He refused it. She lifted him up on her lap. He sniffed once at her plate, then reared against her, trying to lick her face. She held him tighter, smiling over his sleek head at Willy. "He's a problem child," she said.

Willy thought that he was more like a snake, writhing there in the girl's arms. "I reckon he gets pretty nervous, living the way he does," she said. Catherine had turned her head and was gazing out of the window. Willy contemplated her profile. She had always thought that her niece resembled one of the more elegant movie stars. Marlene Dietrich? No, the nose was too aquiline, the jawbone too angular for Marlene. She was more like Greta Garbo, with all that pale hair falling about her shoulders. She found herself speculating on the life lived in that faraway city that she had never visited, probably never would visit. You got glimpses of it sometimes from Catherine's talk. Catherine was always so polite to older people. It did not seem a duty with her. When you were talking together she gave you her whole attention. You two seemed shut in a world of your own. But she knew that Catherine was different when she was with people her own age. Once, Willy, taking a nap in the downstairs bedroom, had been wakened by light, rapid voices on the porch: Catherine and Louise Ellis talking about a woman who had visited Catherine and Jim at a summer camp they had taken in Maine. Louise was laughing because this woman and a man guest, not her husband, had been put in bedrooms so far apart that they could not conveniently spend the night together. "I bet Sarah was furious!" Louise kept saying. Catherine had laughed, too, lightly. "Jim was the one. I wanted to ask them up for Labor Day and he wouldn't let me. Said they wouldn't have enough confidence in me to come."

Willy had never forgotten that snatch of half-heard conversation. In its light she had found herself unconsciously revising her opinion of Jim Chapman. She had always liked him. He was a man, she thought, whom women naturally liked. A tall man who would have been handsome if his features had not been so rugged. His clothes hung loosely on his big frame. A shock of black hair, coarse enough to defy any brush, fell untidily over a wide forehead. His eyes were gray and deep and, when he turned his full gaze on you, piercing. But he had a way of looking at you and not seeming to see you, or of walking, head bent, gazing on the ground. It was the abstracted manner peculiar to scholars, she supposed. Often when you addressed him he did not answer at once and you would think he had not heard you, but she had discovered that he always heard you and invariably answered, sometimes five minutes after you had spoken. Once they had walked to the mail box together and he had asked her a great many questions about the family. She had told him how in the old days two brothers had divided the original tract of land, one taking the tract that had the most white oaks on it and calling it Oak Quarter, and the other, her great-grandfather, a man who liked to fish, choosing the land that had the most water on it so that his farm had taken its name from the long pool which in those days was much used by wild swans. She had even gone back to her great-great-grandfather, the original settler, who had obtained all this land as a Revolutionary grant, a son of old "Irish John Lewis," who had to flee the old country as a young man because he had killed his landlord. She was not one to praise her own family, although the Lewises were good people and had always held a high place in the community, both here and in Virginia. But he seemed so much interested that she had mentioned a half-forgotten family legend. Irish John had been a brave Indian fighter, but in his old age, his dotage, he had exhibited a very different character, cowering in the chimney corner and starting, the story went, every time the door opened, fearing a sheriff had come to arrest him for the murder committed fifty years before.

Chapman had made no comment when she told him the story and she, thinking his interest in the family history exhausted, had changed the subject. They had talked on the way back about Buff Orpingtons. She was raising Plymouth Rocks at the time but was thinking of changing to Buff Orpingtons. But the hens, she had heard, were poor mothers, an undesirable trait when you raised chickens the old-fashioned way and depended on the hens to carry the flock. "Well," he had said, "you must remember it's a circle we're traversing—not necessarily vicious," and smiled and pushed the screen door open. She had thought her niece's husband a little touched in the head until she realized that he was not talking about the hens but about old John Lewis' second childhood.

But scholars, of course, were always absent-minded. There was old Doctor MacIntosh, who lived at the Arlington Hotel in his old age and every morning at ten o'clock would come down and ask the clerk to please tell him whether or not he had eaten breakfast. And when Jim Chapman did give you his full attention, when he turned his piercing gray eyes on you, there was something flattering about it. She had always thought that he had an aristocratic bearing though she couldn't for the life of her tell you anything about his family. He came from some town in the Middle West but he was, according to Cousin Owen Purdue, a scholar and a gentleman. He taught history in a university in New York City but before he became a teacher he had written a long book on Venice, which Cousin Owen said compared favorably with Motley's *Rise of the Dutch Republic*. Her mother, who was of the old school and had severe standards, found him quite correct. Indeed, unless you caught him in one of those fits of abstraction, he was more punctilious than Southern men are nowadays, always holding your chair for you at table or stooping to pick up your handkerchief. She could not imagine his saying anything in the least suggestive in the presence of a lady. It was hard to imagine his condoning the irregularities which the girls had referred to. But Catherine had been explicit. He had not minded the man and woman who were not married staying in the same room while they were in his house and had only been annoyed with Catherine because she had not given them adjacent rooms!

You read about things like that in novels by "The Duchess" and Ouida. That man whom Vere Herbert's mother had married her to had had any amount of mistresses and had been displeased with his wife because she did not take a lover. But she had certainly never expected to encounter such goings on here in America, almost in her own family. People were saying that the times had changed. Perhaps, with all the foreign travel, life over here was getting more the way "The Duchess" and Ouida had pictured it.

She imagined a night club in New York. There was a blare of music and then a girl came out on a stage, no, right out among the people. She had heard Catherine talk about floor shows. The girl wore tights and the ruffles about her hips were sewn with gold sequins. A man got up from a table and danced her about. The music blared louder, until everybody was dancing. All the girls the men were dancing with wore tights and spangled ruffles about their hips and they kept rising in the men's black, clutching arms until suddenly they were whirled away in a vortex of gleaming silk and spangles . . .

The dog jumped to the floor. Catherine pushed her chair a little back from the table. As she turned her head so that she faced Willy, a breeze came in at the window and gently lifted the ends of her long hair. She threw her head back in an instinctive movement and her hair fell away from her temples. She had been wearing her hair that same way the last time she was here. Her figure did not look any more mature now than it had then and yet, Willy thought swiftly, Maria was right. She was too thin, and along with the thinness went a look of fragility that she had not had five years ago. It showed in the fine skin of her temples and in little lines about her mouth. She made a rapid calculation. It seemed absurd but Catherine must be over thirty. In the old days that was middle age—Mama always said that she put on caps the day she was thirty. Nowadays people looked so much younger than their age but perhaps it always showed, if you looked close enough.

"Catherine," she asked suddenly, "did you ever meet the Duchess of Windsor?"

Catherine shook her head. "No," she said and smiled, "I wish I had."

Maria was back with a plate of cakes for Willy and Catherine's coffee. She set the cup down beside Catherine's plate and stood, her head a little on one side, looking down at her intently. Willy suddenly saw Maria as she must appear to a stranger: the black eyes that rarely met yours, the big mouth that always drooped so that you felt that she was sullen even on the days when she was in a good humor. She stirred nervously in her chair. "I'll take some more cakes, Maria."

"They ain't ready," Maria said. She continued to stare at Catherine. Suddenly she laughed. "I been cooking thirty years," she said. "Some mornings I wake up and think about the stomachs got to be filled and I feel like pulling the kivers up over my head and letting the day go past me."

"We don't eat so much, Maria," Willy said mildly, "and you know you have cold supper most any night."

Maria raised her head. Her eyes were brighter than Willy had ever seen them. Her lips smiled contemptuously. "They ain't any sense in eating at all," she said. "If folks wouldn't put nothing in their stomachs for three, four days the Lord would take them away from here."

Catherine was looking up at the old woman. "It takes longer than that, Aunt Maria. Three or four weeks. Some men have lasted longer than that."

Maria stared over their heads at the opposite wall. Her lips moved soundlessly. She sighed. "There was the Saviour. Fasted forty days and forty nights and didn't help him out of none of his sufferings." She picked up the plate of cold cakes and went back to the kitchen.

"I don't know when I've heard Maria so talkative," Willy said.

Catherine laughed. "We're kindred spirits." She crushed her napkin into a ball and put it from her, so roughly that it fell to the floor. As she straightened up after retrieving it

Willy saw that a spasm, as if from secret pain, contracted her delicate mouth. She looked past Willy and spoke, at random, it seemed. "How old is Aunt Maria?"

"Sixty-five, I reckon," Willy said. "Maria hasn't been the same since her last son went to the pen," she added in a low voice. "Seems like it soured her."

"Poor thing," Catherine said absently. "What did he do?"

"Knifed a man. He did some time for that, then killed another man the first day he was home. Maria wanted us to get up a petition to get him out but Mama wouldn't sign. She always said Jesse was no good . . . I think Maria holds it against us."

Catherine ground a cigarette butt out in her saucer. "Aunt Willy . . . why does Cousin Daphne go by her maiden name?"

"She took it back after the divorce."

"Wasn't there some sort of scandal? Didn't they separate on the wedding night or something like that?"

"I don't know," Willy said. "I never asked Daphne and she certainly never told me."

"Mother told me all about it once but I forget . . . Can you imagine any man marrying Cousin Daphne? At least if he was sober . . ."

"From what I hear he was rarely sober," Willy said drily.

"That explains it then. He married her when he was drunk, woke up one morning and saw that face on the pillow. It would be a little like finding you'd gone to bed with a horse! . . . No, a camel. She looks like the camel, after he'd gotten his hump . . ."

Willy glanced at her quickly. Her niece's manner when

she had arrived the night before had seemed to her abrupt and unnatural. Indeed, she had seemed strangely unlike herself all evening. And now this extraordinary remark about Daphne Passavant. "Is she growing coarse?" she thought. "None of the women of our family have ever been coarse, whatever their failings." She stood up. "Daphne'll be down in a minute," she said.

Catherine laughed. "You can't face her across the breakfast table?"

"I've got things to do," Willy said, "and once Daphne gets started on one of her long stories . . ." She broke off, meeting Catherine's ironic gaze. "Daphne's had a hard life," she murmured, "a great many trials. Even before her marriage . . ."

"It looks like everybody around here has a hard life," Catherine said, "except you." A charming smile came on her lips. She laid her hand on Willy's arm. "You just have a hell of a good time, don't you, Aunt Willy?"

Willy started and moved away from her, out into the kitchen and from there to the porch that ran the length of the ell. She turned around. Catherine was beside her. "You want to come with me while I feed the chickens?" she asked.

"Yes," Catherine said, "let me come with you."

Willy looked down at the dog, who stood, wagging his tail and gazing with bright eyes at Catherine. "We better not take him," she said. "He might run the chickens . . . Maria, you keep him up till we come back . . ."

Maria had sat down on a stool at the end of the kitchen table and was eating her breakfast. She gazed straight before her, chewing steadily. "He can stay in here if he behaves himself," she mumbled.

"Mother'll be back in just a minute, darling," Catherine said and closed the kitchen door behind her. Willy took a tin measure and, scooping grain from a sack, poured it into a bucket. "You might just bring that," she said, indicating another bucket full of water that stood on a table.

Catherine picked up the bucket of water and followed her across the lawn to the chicken yard. A wire fence had once encircled the yard but it was so weighted with honeysuckle that it had long ago fallen to the ground. Saplings had grown up along the line of the fence and the vines clambering up them had erected themselves into a wall higher than the original fence. They passed between the green walls into the enclosure. Catherine cried out when she saw the white birds dotting all the ground. "Aunt Willy! Did you raise them all?"

"White Leghorns," Willy said. "I started out with two hundred baby chicks."

She began scattering handfuls of grain on the ground. "You might just pour that water there," she said, indicating a round tin pan that stood under the drip of a log trough. Catherine poured the water out and walked a little way up the hill, examining the trough.

"The water doesn't run from the spring any more?" she asked.

Willy came over and, standing beside her, looked down at the green, rotting wood. "I haven't cleaned the moss out in a long time," she said. "I ought to get a pipe up to the spring. But there's so much to do . . ."

The two women stood silent while the chickens crowded around. A breeze stirred the leaves over their heads. Willy, lifting her eyes to the woods, saw that all the leaves had a new-washed look. And the sunshine seemed almost liquid. The shadows that spangled the chickens' white bodies quivered. The yellow legs might have been moving through a stream.

A young rooster cocked a red-rimmed eye at her and crowed. She flung a handful of grain that spattered off his sleek coat, then let her eyes rove over the flock, counting swiftly. One hundred and twenty-eight, if she counted right. She sighed. It was pleasant to be out here in the cool air satisfying the needs of these handsome creatures, pleasant to have a companion as you went about your morning tasks. She could still feel the light touch of the girl's hand on her arm. It had seemed to betoken a spontaneous affection. The memory warmed her. She had never thought of Catherine or anybody else as being particularly fond of her, except, of course, Mama and Cousin Daphne, who depended on her for so many things.

She flung a last handful of grain, then dusted one hand against the other. She smiled at Catherine. "You want to see something?"

Catherine nodded. They left the chicken yard and took the path to the stable lot.

Once Catherine stopped and looked back at the house. "Aunt Willy, how old is Swan Quarter?"

"Your great-great-grandfather built it," Willy said, "the main part, that is. It's been added on to so many times it's hard to tell how old it is." She stopped, too, and looked back at the house where it lay, gray and squat under the high oak boughs. "I thought about having it painted a few years ago. Had a man come out from town. He figured it would cost two hundred dollars just to paint the house, not counting the ell and the wing."

"I like it gray," Catherine said. "It's been gray ever since I've known it. In the old days, in New England, they didn't paint houses, Aunt Willy. Just let them weather."

"That so?" Willy said. " 'Too poor to paint, too proud to whitewash,' I reckon. I thought about whitewashing once but Mama had a fit and I gave the idea up."

They came to the gate. She unfastened it. They passed through the dark runway of the stable out into the bright sunshine of the lot. In the shade of a maple tree a burly, middle-aged man sat on the top rail of the fence. Before him, nosing at his blue jeans pants leg and whinnying a little, stood a shining, dark red horse. As they approached Catherine saw that he was a young stallion.

He heard them and turned around. For a moment he hesitated, then pressed nearer to the man, nibbling at his pockets, jerking his head up, holding his quivering lips against his face.

The man suddenly brought his right hand out from behind his back. The horse snatched the apple he had been holding and demolished it in one swift crunch and moved over and stood beside Willy, nudging her elbow.

She put her arm about his neck. "Morning, Mr. Shannon," she said. "This is my niece, Mrs. Chapman. You remember Agnes?"

Mr. Shannon jerked his broad-brimmed hat off, then

settled it firmly on the back of his head. "I sure do," he said, "and I remember you, too, Miss, when you warn't no higher than that." He had jumped from the fence and stood beside them, holding his sun-burned hand level with his knee. His blue eyes, which looked light in his red, grizzled face, met Catherine's eyes. He smiled.

She smiled back faintly. Her eyes went to the stallion. "My *God,* Aunt Willy!" she said softly.

Shannon laughed. He broke a switch from the bough over his head and, going over, caught the stallion by his fore-top, then flicked his breast with the switch. The stallion tossed his head and, backing a few steps, took his stance as if he were in the ring. The sun glistened on his coat that was everywhere dark red except for a narrow white blaze on the forehead and one white hind foot.

"My God!" Catherine said again.

"I don't reckon you ever seen one of these before," Shannon said. "It's a new breed. The Tennessee Walking Horse . . . *Hi-yuh, Red*!"

He stepped back and the stallion began to trot around the enclosure. Shannon called to him until he increased his pace, then halted him abruptly. He bent and, taking hold of a hind hoof, indicated the print it had left. "See how he overreaches? That's the Allen walk. Comes straight down from old Black Allen."

"It looks sort of like a fox trot," Catherine said.

"It's a saddle gait, all right, but it's faster'n any other horse can saddle. We call it 'the plantation walk.'" He straightened up, smiling. "Miss Willy's the leading breeder in this part of the country."

Willy blushed. "He's the first horse I ever owned in my life!"

"You wouldn't a had him but for that old nigger of yours," Shannon said. He looked at Catherine. "The Manigaults bought his mammy from Red Gower over in Lewisburg. Then Tom Manigault came down with pneumonia two days before the colt was foalded. Boar got loose in the stable and mare stepped on her colt and broke its leg. Miss Willy's nigger, old Joe, was working over at Manigaults' about that time and Miss Elsie Manny, she told him to make 'way with the colt.

"Joe took an axe out in the pasture and then he come back, says, 'Mis Maney, that colt look at me so pretty I can't kill him.' 'Well, take him off,' Mis Manny tells him, 'I don't care what you do with him, long as you get him out of here.' Joe, he slung that colt in a sack and brought him right over to Miss Willy, knowing she was a good hand with sick things."

"And I ran right up the hill to you," Willy said. "I thought I'd die when Joe brought that colt in with his leg all dangling."

"Did you set it?" Catherine asked.

"Miss Willy was for putting on a splint but I said let Nature take its course. A young thing like that, its bones ain't nothing but jelly. Miss Willy, she didn't like to ask the Manigaults to send the mare over so old Joe he'd go over twice a day and milk her and Miss Willy'd feed that colt out of a bottle. She kept him in the old wash-house till he got so stout he most tore it down."

Catherine walked over and examined the stallion's legs. "I can't tell which one it was," she said.

Mr. Shannon shook his head. "There ain't a blemish on him," he said. "Miss Willy, you going to sow that alfalfa?"

Willy nodded. "I thought I'd buy some seed from Tom Manigault—if you think the price is right."

"I don't reckon you can get it any cheaper—now," Mr. Shannon said. "You could a saved a cent if you'd bought two weeks ago. Seed's going to be high this fall."

"I'll send him word today," Willy said. "Thirty pounds'll about sow the upper bottom, don't you think?"

"Let me take the message, Aunt Willy," Catherine said suddenly. "I can ride Red over."

Willy shook her head. "I don't think you'd better ride him."

"She'll be all right on Red," Mr. Shannon said easily, "if she's just going to Manigaults'. All their mares are over on the far pasture and these days you don't meet no horses on the road. Red's all right, long as he don't git around other horses."

"Yes," Catherine said decisively, "he'll be all right."

Mr. Shannon yawned. "I better be getting back to the house." He turned away, then halted. "Miss Willy, the fair'll open in two weeks."

"I don't ever go," Willy said.

"You ought to go this year," Shannon told her. He jerked a thumb at the horse. "You take Red over there, he'll win, in any class you show him. They ain't nothing in this country can touch him."

Willy laughed. "You talk like I was Elsie Manigault, Mr. Shannon."

He laughed, too, then, pushing his hat farther back on his head, chewed at a grass stem for a moment. "It'd cost you something to haul him over there," he said, "but you'd make it back in stud fees. Red's ready to *stand* now, Miss Willy. But you better show him this year before he stands. A stallion just don't show as well after you start breeding him. Gets coarse looking."

"I—why, I couldn't leave Mama," Willy said.

Catherine went to her suddenly and put her hand on her arm. "Yes, you can. I'll take care of Mammy. Sure enough, Aunt Willy. You ought to go."

Willy looked over at Red. He was standing with his head lowered, greenish juice from the apple still trickling from his lips. He suddenly jerked his head up and with a little fling of his hind quarters made off for the other end of the lot. His coat, she thought, was the true chestnut, the exact shining red of a nut fresh from the burr. His dark glance as he turned away just then had been full of affection. He had good manners—Mr. Shannon had seen to that. It would be no trouble to show him, if you kept your mind on the horse and didn't notice all the curious eyes looking down at you.

Elsie Manigault showed her horses all over the country. In the fall you couldn't pick up a paper without seeing Elsie and one of her horses. In those pictures she always seemed to stand more proudly than usual, her head held as high as the horse's, and her rather impassive face touched with an eager look that it did not ordinarily have. For a moment she saw herself in Elsie's place. At night. In the ring. There was a big crowd. She had just brought Red in. He stood still. The lights fell full on him. The people's faces all seemed to draw nearer and all of them were kind.

She smiled up at them. She knew what they were feeling. It was a long time since she, a girl of fifteen, used to go with her brother to horse shows around the country. It was always hot and dusty. You would wish you had not come and then a horse would walk into the ring and as his trainer turned him slowly around, the light would strike on him and you would get that feeling of sudden, almost uncontrollable excitement that you never got at any other time, that she had never got since—and then over only a few horses. It would be like that if she showed Red.

Catherine was still pressing her arm. "Sure enough, Aunt Willy."

Willy gave a little laugh. "I don't know," she said. "I *might*—if everything works out right."

III

On the way to the house Catherine suddenly halted. She said: "I believe I'll walk up to the spring."

"All right," Willy said, "yonder's the path."

Catherine started up the slope, walking fast as long as she was in sight of the house. Once she was in the woods she went more slowly.

She came to the spring: a pool at the foot of a sycamore, walled in by ancient, crumbling stones. On the slope above young beeches grew thick. A path curved up between limestone boulders. The beeches grew sparser. Two huge boulders seemed to block the path, then showed a space between them a little larger than her body. She slipped through the aperture into an open glade.

In the middle of the glade a beech tree had grown up uncompanioned. Its boughs, on which a few of the leaves were already a pale yellow, swept the ground. In front of it a flat rock was embedded in the earth. Ferns thrust up against its sides. Here and there the stone was crusted with bluish, fan-shaped lichens. She gazed at the tree which was not much older than she herself was. When she was a child and came here alone to play she used to fancy that the beech was a woman, advancing with wide-spread skirts. The flat rock was the woman's table. The child would take a sharp stone and sprawling beside it laboriously scrape off the lichens in order that the table might be fair for the cloth that the woman might spread. That would have been when she was eight years old. She would be thirty-five in November; the woman was not much taller than she had been then; and the same lichens still grew on this rock.

She stepped backward into a drift of last year's leaves, felt them brush damp against her bare ankles, took a step forward and flung herself on the ground, shuddering all over.

A maple seedling, caught in the rush of her passage, slipped out from under a fold of her skirt, trembled a moment, then stood erect, its tap root still firm in the rich mould of decomposing leaves. The dry leaves on top of the drift rustled under breathing. There was no other sound in the wood.

Her shudders ceased. She sat up, brushing the back of her hand across her face.

Nobody shall ever know what happened to me in that room. I came in. There was a letter on my dressing table. From Roberta Ellis. She wanted us to come up to Westport. But I saw the other letter fallen on the rug and I stooped and picked it up. The time after that all seems so long but I remember walking in and laying it on his desk. I wish I had brought it. I want to read it again. But the words do not change.

> *Dearest: The Pennsylvania at two-forty. My brown gloves are in your overcoat pocket. Please remember to bring them and the Brahms record if you go by the apartment. Yes.*

I thought: the Pennsylvania. *He* said *the Grand Central: New Haven; Bill Dyer: about the history text-book. I looked in the cabinet. The Brahms record was gone.* Brown *gloves. I opened his closet door and then I shut it. I did not want to touch his clothes. The note was in pencil on a scrap of paper, the signature some sort of symbol.* Yes *means Yes, I love you. In answer to a question.*

For a minute I did not know who she was and then she came into the room and stood beside my bed and smiled and said, Yes, I am the one. *She called me* Mrs. Chapman, *the way she did the only time I ever saw her, when Bob Tilman brought her here for dinner. She had on a black dress with a heliotrope belt. She is taller than I am and her hair is black and naturally curly. She wears it quite short. Her eyes are blue. She did not talk at all that evening, only her eyes kept going from face to face. I thought: What keeps you from being happy? Maybe you'd like to get married . . .*

Yes *means* "Yes, I love you." *In answer to a question. Why did I not know then that she was the one? That night, after everybody had gone we talked about her. He said she was the only woman in the department who was worth a damn. He had just opened the window and was coming back to bed and the cool air flowed into the room and over my bare arms, folded outside the coverlet. I was slipping down between the warm sheets and I turned my head and said "How? How is she good, Jim?"*

He said: "That piece on Mazzini. Badly organized, but she picks the significant episode." He was standing in the middle of the room and he yawned and stretched his arms over his head and I thought: How coarse his black hair is and I must not let him buy any more striped pajamas, for they make him look too lean and then he was beside the bed and his hand was going out to turn off the lamp and he looked down at me.

You wade deep in time but you cannot feel its lapping against breast or thigh or knee. The stream is colorless until some wave, striking on rock, shows an edge of foam. He has a long face, a little sunken in the cheeks, an obstinate chin. His eyes are too pale a gray, except when the pupil, enlarging, turns the eye black. I used to wait for a look he had. It was one of surprise, as if he had not known that I would be there. Sometimes his lips, opening, would seem to tremble. That night he looked at me and I thought, "It has been a long time," and then he bent and set his lips against my arm, the soft part where it is joined on to the body, and I was about to move over in the bed when suddenly his lips went away and the light clicked off and I lay there in the dark, wondering, and then I heard him turn over in bed, heard the thump his hand always makes when it strikes against the head-board and I knew he was gone away from me. Into sleep, I thought. I did not know that it was forever.

She began to weep, leaning back against the rock, her naked face up-tilted. A warm tear was salt in her mouth. She parted her lips, breathing hard like a child. The tears ceased. She sat quiet for a while, then picked up a stick and listlessly made marks on the ground.

It had been three days since she had discovered her husband's infidelity. The time seemed longer than all the rest of her life put together. There had been first the terrible quiet. You looked out on the street and you could have pushed the houses over with your hand and the cars and the people did not move. If you could have stood there while it flowed around you, flowed over you, stopped your eyes, your ears, but you knew that at any second sound would come like thunder, and then you knew that it was your own least movement that would set it crashing all about you and, like an acrobat, hurling himself against the brittle, flaming hoop, you stretched out the hand for the letter.

And having moved you could not stop, but must go on faster and faster, until once there in the darkness it seemed that the machine could not cling longer to the round of earth but, tilting upward between the leaning trees, would lose itself in the grayness to which as you drove you kept raising your eyes. . . . But that was a long time ago, she thought. And besides, the wench is dead. Or if she is not dead there is very little of her left in me and I wish that little were out of me. She looked up at the sky, visible only as patches of intense blue, pricked everywhere with green. I would like not to be, she thought. That is why I came here, where there is not much living.

Yes *means* "Yes, I love you." *In answer to a question.* I will never tell him, she thought. I laid the letter there on his desk, but I did not write anything on it. I do not often go into the study. He will be glad to find the letter and will think he has misplaced it and burn it. Later, when I can think, I will tell him I want a divorce and will make up a story that will deceive even him. That will be my revenge.

She pushed a drift of leaves aside with the stick and the earth showed, black and gleaming. Revenge is what people want when they hate each other, when they want to kill each other . . . I must not ask myself again how he came to betray me. I do not believe he is in love with that girl. It is his way of striking at me, as if we were standing on a precipice and he put out his hand and pushed me over. But he must have thought what it would be like after I was dead! Children, who are savages until they are thirteen, sometimes kill each other in play. Does the desire live on, submerged, in every human being? Does everybody, at bottom, hate everybody else? That Russian somebody brought to the apartment once. He and his companion, lost in Siberia, had to walk for days within calling distance but out of sight of each other. "But why?" I asked. "Did you hate each other so?" "We were the only ones," he said. Do lovers, forever traversing a wide waste in which they are the only living beings, always come to hate each other?

Leaves rustled. Somebody was coming up the path. She got to her feet, smoothing her wrinkled skirt. Blue checks showed between the rocks. A black-shod foot, a lean, lisle-clad leg appeared. Cousin Daphne wriggled through the opening and stood facing Catherine. She wore a sun-hat and carried a basket on her arm. She looked at Catherine. Her black eyes glistened. She stepped forward until her sallow face was so close that Catherine could smell her warm, sweetish breath. She was, incredibly, whispering. "You get used to it. I didn't sleep. For a year. I took things. Doctor Watson thought I was forming a habit . . ."

She stroked Catherine's shoulder. Catherine, turning her head a little, could see the fingers, the nails blunt and ornamented with purplish polish, digging a little into the sleeve as they slowly descended her arm.

"They'll come on down," she thought. "She'll take hold of my heart."

She stood back on her heels. Her body contracted all over.

If, a moment ago, she had only taken time to wipe her face on her skirt! But her cheeks felt dry. If there was any moisture on them it could be sweat. It was hot in the woods.

She took the cool, limp hand in hers, detached it from her shoulder. She spoke, with fury, in a low, harsh tone. "What is the matter, Cousin Daphne?"

Daphne stared at her with bright eyes. Color came in her sallow cheeks. Her lips trembled. "I thought you might be in trouble . . . seeing you off here by yourself . . ."

Catherine felt her own eyes as hard, as impenetrable as marble. "I like to get off in the woods," she said. "I used to come up here to play when I was little." She put her hand out, touched Daphne's basket. "What have you got?"

"Mushrooms." She gave Catherine a quick glance. "It is my hobby. I know two hundred and three varieties." She laughed breathlessly. "The Tricholomae and Cortinarii have always confused me. I hope to master them this summer."

"How will you go about it?"

"Oh, I have books, and correspondents all over the country. We mycologists—I am also, I may say, a mycophagist; not all mycologists are—we send each other specimens. In that way we can add to our knowledge."

"What in the hell is a mycophagist?" Catherine asked.

"One who eats mushrooms," Daphne said and gave a little shriek and stooped. "A *Phalloides!*"

"Are you going to mycophagize it?"

"It's poisonous," Daphne said. "The Amanita Phalloides. See. There's the Death Cup."

Catherine took the mushroom from her, turning it between her fingers. The knob-shaped surface was white everywhere except for a faint stain of green in the center. She turned the mushroom upside down. The underside was like accordion-plaited white silk. The stem was dead-white, ornamented with a fragile, silky petticoat and ending in a cup-shaped sheath. The Death Cup—*volva* Cousin Daphne was calling it now—was even more delicate than the rest, glistening like the petals of a white tulip that has been kept in water too long. *Galen,* Cousin Daphne was saying. . . . *Pliny . . . the breath of serpents.* Her own fingers were bruising the stalk that in all countries, all ages, had struggled through crusted earth to hold this cup up to the reaching hand. *It was all right at first. I was numb. But the pain is a fire now. And all the time the mind running like wind and flames leaping up from embers that I thought were dead . . .*

She looked overhead. The sun was high but only a few gold disks of light struck through to the rocks. Under the low boughs of the "lady" beech a few fallen yellow leaves shone with dew. Even at mid-day this place kept its secret look. The mushroom whirled slower between her cold fingers. In a few minutes Cousin Daphne would go away . . .

" . . . You'd better wash your hands. You might put them in your mouth."

Catherine looked down at her hands, wet from the bruised stalk. "Does it take very long, to die from it I mean?"

"Several days. That is the strange thing about Amanita poisoning. The symptoms do not appear for ten or twelve hours. Then there is nausea and vomiting, followed by convulsions and finally death. Captain McIlvaine has a very interesting account of two deaths from Amanita poisoning in his 'One Thousand Varieties of Mushrooms.' Would you like to read it?"

"No," Catherine said absently.

Daphne's lips quivered in a nervous smile. She picked up the basket she had set down. "I'm going over here a little way to hunt *boleti,*" she said. "You want to come along?"

Catherine looked off through the leaves. She could see down into the stable lot. The young stallion moved about there, cropping the short grass that grew in the fence corners. The sun shone on his red back. He suddenly cantered up to the fence and stood with forward pricked ears, gazing off over the fields.

She turned and met the restless black eyes. She shook her head. "I'd love to, Cousin Daphne. But I've got to ride over to Manigaults', to take a message for Aunt Will."

IV

THE HORSE MOVED nearer the gate, nudging it shut with his nose. Catherine bent from the saddle and let the chain fall back on the nail and sat looking down the slope. The bottoms east of the house were yellow—most of the crops had been cut—except at the far end where they were rimmed with green. The woods were thick; only in one place could you glimpse the little river that, flowing in a wide bend, encircled all of the land except this wooded hillside here to the north.

Her eyes came back to the house. The trees grew so close about it that you could not discern its shape. Here and there a clapboard shone silver through the leaves. She closed her eyes, overcome by recollection. Once, long ago, she had sat on this hill and watched the sun strike that same shimmer out of gray wood. She had been on horseback then, too, in the flat, English saddle in front of Uncle Jack. It was the day her mother would not let her go over to play with the Robinson children. She had cried so hard that she had been sent from the table. She had stayed out in the hammock and fallen asleep. When she woke Uncle Jack was looking down at her. He bent and picked her up, calling out to her mother, sewing at an upstairs window, that he was going over to Duncan's shop to see about a mower blade. His horse was hitched at the stile. When they got to the top of the hill—there was no gate here then—he pulled up and they looked back at the house. The sun slid through the leaves and struck the same shimmer out of the gray clapboards. A tiny figure had come around the corner of the house: Maria on her way to feed the chickens. They had watched her pass across the yard and out of sight behind the shrubbery. Uncle Jack laughed and she, five years old at the time—no, perhaps it was the summer she was six—turned and tried to hug him, and he laughed again and laying the flat of his hand against the small of her back pressed her up against his warm body. Poor Jack! A scapegrace, perhaps downright dishonest—there had been some sort of funny business about the mortgage—but he had a charming kindness for children, always wanting them to have the things on which their hearts were set. Was it because his own desires were so often frowned upon—he must at that moment have been involved in the slightly disgraceful affair with the Reynolds woman—or did he know instinctively how short life can be? He was only thirty-four when he was killed, breaking a horse for a friend.

The stallion arched his neck impatiently and took a step backward. She wheeled and rode along the lane. In all the years she had been away nothing had changed. Blackberry bushes thick on each side of the road and the dust soft under the horse's hooves. That white cottage, set under clipped maple boughs, with the big red barn rising behind it, was the Shannon place. She had used to come here often as a child, with messages for Mrs. Shannon. Mrs. Shannon had been dead five years but the hollyhocks and Princess Feather still grew luxuriantly about the house. Possibly Mr. Shannon kept the same flowers growing in honor of his wife's memory. She wondered if he cultivated them himself or brought a negro man in from the field . . . Mr. Shannon had broken this horse and had spent a lot of time since then training him. Evidently he advised Aunt Will about all her farm matters. They had had a companionable air as they talked there together yesterday. She suddenly turned in the saddle and looked back at the white cottage. No. It wasn't possible. He was a widower and Aunt Will was an old maid and they lived within a stone's throw of each other, but still, down at heel as she and Swan Quarter, too, were—she was a Lewis and Mr. Shannon, for all his white paint and his fine, new barn, was a Shannon. Quentin Durward Shannon. The old lady had explained about his name, last night, at supper. "His father was reading *Quentin Durward* the night he was born." Aunt Willy, eating butter beans, did not look up. Mammy had looked over at her, though, malice in her faded blue eyes. "Quent can read too. He got as far as the eighth grade."

The stallion was pulling at the bit. She tightened the reins. He broke into a canter. You ought not to gallop a horse in hot weather, in fly time, but it was shady along here and besides he needed exercise. It was only a quarter of a mile to the hard road. Would she be able to pull him up in time? If she didn't he might slip and break his leg. He was a stallion. Suppose he got wind of the Manigault mares? Mr. Shannon didn't know. They might not all be in the far pasture. His canter was hard, but that fast walk, the "Allen walk," was like motion in a dream. There is nothing like having a good horse between your legs. It had been so long since she had ridden that she had almost forgotten how it felt. The macadam showed, a dark streak between the leafy shoulders of the lane. She pulled the horse to a walk, halted him and, reaching down, felt under the saddle blanket. He was wet, all right, and she herself was breathing hard. An odor that she loved came up to her, clean horse-hide mingled with saddle leather. She breathed it in and bent and pressed her cheek against his wet neck. He started and reared. She brought him down with a stroke of her crop, let him prance a few steps, then put him into a walk again.

The road went up hill for a little way and then she was on top of the rise. On her left dark beech woods bounded the southern end of Swan Quarter. On the other side of the road the land stretched away in wide, open fields. All this land belonged to the Manigaults. They were rich—city money—and had not had to sell any of their land, like the Lewises. When she and Jim were first married they used sometimes to go out to Cousin Edward Manigault's place on Long Island. And then Cousin Ed had died and his widow—it was her money, not his—had sold the place and gone off to France to live. She had had to come back. Everybody was coming back nowadays, rats deserting the sinking ship, Jim said, and was down here now, with Tom. A restless, handsome woman who had been everywhere and knew everybody. She probably wouldn't stay long. She checked the horse and bringing one leg up over the pommel of her saddle sat at ease, studying the landscape. That first dark mass of trees concealed the Manigault house. Scattered over the sun-lit fields were other dark groves. Old Cousin Lucius had left six when the land was first cleared, for his children to build in. But there had never been but the one dwelling house on the place—Big Pond it was called. She looked past the last grove to the dark line of trees that in this country rimmed every horizon. Somewhere between here and those woods was a chain of ponds where men went in the fall to shoot wild ducks. She had walked there often, with Cousin Robert, Cousin Ed's brother, who until his death a few years ago had lived here and farmed the place.

It was strange how the geography of this country stayed in her mind. Those ponds, a hill at Swan Quarter that was part of the boundary of the farm and was named in the grant: French Station, a spring, twenty miles from here that was the source of Swan Creek. These remote, rarely glimpsed places had for her a reality, an importance that no other places had. She did not think about them often but when she did the thought seemed part of something that had gone before, not something that was starting, as if some corner of her mind had all the time been keeping them in contemplation.

The horse went slower. His head jerked from side to side. He cropped leaves from the roadside bushes, tender leaves, vernal rather than autumnal, some mitten-shaped, others three-pronged, growing on the same pale, brown-spotted stalks.

Sassafras, she thought. I was going to gather a lot next time I was in the woods and dry them for *filé* powder. She tore off a handful, crushed them and held them to her face and breathed in the aromatic fragrance. I could gather some now, she thought. No, I'll get a paper bag at the Manigaults' and fill it on the way home. But she knew that she would not gather them when she came that way again and unclosed her fingers and let the leaves drift from her open palm down on to the road.

But I am all right, she thought. I woke up this morning and knew that I was all right. It is a numb feeling. No, it is like floating, only I am lighter than ordinary. That is because something has gone out of me and lies sunken under this water. If I move carefully it will never rise but will lie there always under the light water. That is the way it will be from now on. That is the way it is with most people.

And why did you ever think it would be different with you?

Because it was. When we were first married. I never had any doubts. I never stopped to ask myself any questions.

Perhaps it would have been better if you had.

It was the same with him.

How do you know?

I always knew. I know it now. If he himself were to deny it I wouldn't believe him.

But he has changed. If you had not been vain and thoughtless you would have known the moment when he changed.

It was last spring, at that party at the Wallaces'. We played that game where you say what flower a person reminds you of or what perfume or what age he or she ought to have lived in or whom he or she ought to have married. Joe Wallace and Jim were sitting on the sofa. Joe looked at me and laughed and wrote a name and showed it to Jim. Jim shook his head. From across the room I could hear what he said: *The Black Douglas. Women like that always want a Black Douglas.* I wrote down Georges Sand because he detests her and then I looked at him. *I am not this kind of woman or that kind of woman. I am Catherine. There is not anybody else like me. You know that.* He smiled but his eyes did not soften when they looked into mine. I looked away.

There were other times: the week Ellen Page was visiting us. I was kneeling on the hearth, doing her hair a new way, and he was leaving to go to that girl's apartment, to work on some papers, and I laughed and said how attractive she was, but I did not think she was attractive, I thought she was a poor thing. It happened then, that night, or had it already happened? Had they been meeting for weeks, months? There is no way I can find out. Even if I see him again, talk with him. Or if I went to her, threatened her, brought her to her knees. . . . *I am aware . . . my husband. . . . You have been having an affair with my husband. . . .* No, I would not go like that. . . . Miss Ross *. . . to plead my husband's cause . . . such an excellent arrangement . . . so much in common . . . anything . . . to facilitate . . . Reno? Yes, next week. Or perhaps Arkansas. I believe it is even quicker. . . .*

She let go the reins and leaning forward, her eyes shut, rested her cheek against the stallion's red, shining back. Startled, he plunged to one side, then broke into his swift walk. She sat up. They surmounted the crest of the hill, rounded a bend. She pulled the horse up short and sat, staring.

It is because they were the background of your childhood.

She brought her leg over the pommel of the saddle and thrust her foot deep into the stirrup. "You stay away," she whispered behind set lips. "I'll be all right. If you'll stay away."

The entrance to Big Pond was only a few feet ahead, beyond that sycamore tree. She rounded the bend and stopped, amazed. She had awaited an unpainted, five-barred gate, an avenue of scraggy cedars. But here were brick pillars, white against green mounds of box. She turned in on a gravelled drive, looking about her curiously as she rode. The althea circle she had known as a child was still here. It looked unfamiliar, set off by smooth turf. And there was the sugar tree in whose shade Cousin Robert used to sit every day in summer to read.

A man was reclining under it now in a lawn chair. A dark-haired, middle-aged man. He wore nothing but a pair of spectacles and a pair of mulberry-colored shorts. He heard the horse's hooves crunch on the gravel and raised his head. As horse and rider approached he closed his book, marking the place with his finger. When they came opposite the chair he got up. The paunch that was not visible when he was reclining distended the waist-band of his shorts and shook a little as he walked towards her on thin, sun-burned legs. He was smiling. He said: "I am LeRoy Miller. You are Mrs. Chapman, Miss Willy Lewis' niece."

She reached down a hand. "How did you know who I was?"

"The grape vine. Rodney, old Joe's worthless son, comes over every morning and pretends to sweep out my room." His eyes were dark between heavy lids. He let her hand go after holding it an instant too long. He was smiling again. "You live in New York and have come south for your health. You are pretty puny but Miss Willy will build you up. With buttermilk." His voice suddenly rose to a falsetto.

" 'T'ain't no use, though. You'll go back to New York and git to running around nights and it'll all to be done over again . . ."

She did not answer, staring past him at massive white columns, a deep, recessed porch, wings so wide that they disappeared behind the evergreens. Where was the old red-brick house with the two square turrets? It had been ugly but sound. Were its walls concealed behind this gleaming façade or had the Manigaults razed them before they erected the new house?

The man's dark eyes brightened. He took a step forward. "You haven't been here since Elsie's improvements?"

She shook her head abstractedly and rode a little nearer to the house, still staring. He followed. He made a gesture with his hand. "I know it's banal. But it was a choice. Mount Vernon or Westover. Westover isn't spacious enough for Elsie. You know Elsie? Besides the countryside cries for Mount Vernon. We decided, Elsie and I, that it should have it."

She looked down at him. One corner of her mouth lifted a little. "You might have given it a Norman farm-house."

"Oh, come," he said, "I'm a reputable architect." His eyes went back to the house, dwelt on it a minute, then returned to her face. "Are your objections aesthetic or sentimental?"

She felt an impulse to dissemble. "Oh, I like it. Immensely. It just gave me a turn, coming on it, sudden-like. I used to visit here when I was a child," she added hurriedly. "In summer Cousin Robert always sat out under that tree to read . . ."

He turned and looked at the tree. "And now . . . the wild ass . . . sits in his chair and does not break his sleep." He grinned. "I'm not so wild." He was looking straight at her, the heavy lids dark and shining as if they had been burnished, the black glance liquid yet febrile. You know me for what I am, it said, so why any concealment between us? See? I can joke about it.

She looked away, gathering up her bridle rein. "Is Cousin Elsie at home?"

"Elsie got up at dawn, to valet a mare. She's probably still at the stable."

"Well, Tom is the one I want to see. I have a message from my aunt."

"Tom is out in the field, combining clover."

She laughed. "With what?"

"With air. He has an infernal machine that sifts grain from chaff." The horse was moving on. He walked a few steps beside him. "Would you like me to show you the way?"

"Thanks. I'll just ride around to the barn."

He nodded, turning away, taking his dismissal with an almost child-like alacrity. He was stretching himself out in the bright-colored chair. Over his head the broad sugar-tree leaves hung motionless, as thick, as coarsely green as they had hung in her childhood. What was he doing here in this warm, unstirring air? Had the city spewed him and his kind out so that they would roam even these remote fields?

The lane she was riding down was bordered on each side by a neat running-board fence. The stable came in sight:

not the old double crib log stable of Cousin Robert's day but a white, clapboarded affair. The weather vane was a running horse. Off to the side was a paddock. Aunt Willy had told her that Elsie Manigault was raising walking horses. Why had she omitted to mention the renovation of the house?

A negro boy was lounging in the stable runway. "You seen Mister Tom anywhere around?" she enquired.

He jerked a thumb towards the field. "He in there," he said and, slouching across the road, opened a gate. She passed through. Two negro men lay full length on the grass at the edge of the field. A white man squatted on his haunches beside them. He was rising, dark and short, as bandy-legged as a jockey, with a big beak of a nose. Tom Manigault had been fourteen or fifteen years old the last time she saw him: a towheaded boy with blue eyes.

"I'm looking for Mr. Manigault," she said.

"He's coming now," he told her and pointed down the field.

Two machines were approaching, metal bumble bees hooked together, the little one pulling the big one. The driver of the first machine was a negro but there were two white men on the second machine. One sat with his back turned, bending over some sacks. The other stood, making downward motions with his hand, shouting. A lever clanked. Laden sacks shot out on to the ground. The machine rumbled to a stop. The man swung off a platform and stood to one side. "Blow her out now," he yelled. The bumble bees vibrated hoarsely. Brown chaff spilled out on the ground. He stooped and sifted some through his fingers.

"There s too much draft," he said angrily. "I told you all along there was too much draft, Ike."

The negro called "Yas, sir" cheerfully and leaped to the ground. The young man had not seen her. He had stopped where a bucket sat under some bushes. The bright arc rose above his head. Water spilled down, pearling his red face, lodging in bright drops on his thick, fair hair, his little moustache. He strode forward, shaking himself like a dog, talking to no one in particular. "If they could just get one thing right around here. First thing I did this morning was to set that draft and then some fool had to come along and monkey with it . . ."

He saw her and stopped. He looked at the bow-legged man as if he were somehow responsible for her presence, then back at her. He said, "Good morning," and came towards her.

She realized that she must identify herself. She said as she leaned down to shake hands: "Tom, this is Catherine Chapman, Cousin Willy's niece . . . I haven't seen you since you were fifteen years old . . . We went out to the Big Pond, after ducks . . ."

"I don't remember," he said. He kept his eyes fastened on hers. They were gray-blue and in contrast with his red cheeks had an odd brilliance, the iris seeming to be composed of innumerable, tiny, almost metallically shining particles. "I mean I remember that blind we had out there," he added and coming closer took hold of her stirrup and twisted it in its leather thong.

His proximity gave her a shock. She felt that if the other men had not been in the field he would have laid his hand on her foot, her leg. She straightened herself in the saddle.

"Aunt Willy wants to know if you can sell her some alfalfa seed?"

"I'll send some over this afternoon," he said.

She became aware that the little bow-legged man, standing beside her, had all this time been eyeing the horse. He came up to her now. "Miss, can I take a turn on him?"

She said, "Certainly," and dismounted.

Tom Manigault laughed. "Want to ride a real walking horse, Joe?"

Joe grinned and, mounting, rode off over the field. Catherine turned to Tom. "Aunt Willy wants to know the price. She thinks thirty pounds will be enough."

"It's selling for twenty cents," he said absently. He was looking past her, towards the gate. "Here comes Mother," he said. A negro boy was holding the gate open while a woman passed through. She came towards them lithely over the stubble, tall and white-haired, in straw-colored slacks and a vivid green jacket.

Catherine went to meet her. She put out her hand but Mrs. Manigault, disregarding it, set both hands on her shoulders, and kissed her on each cheek in the French manner.

"I came by the house," she said, "and Roy told me you were here. My *dear*! I'm so glad . . . You'll stay for lunch?"

Her hair had grown white—or been bleached—since Catherine had seen her but the full, deep voice had not changed. At this moment it was warm and eager.

As Catherine hesitated she put an arm lightly over her

shoulder. "You must. Tom and I *never* have any company . . . I'll send a boy . . . Tom, they don't have a telephone at Swan Quarter, do they?"

"Yes," Tom said, "they have a telephone."

He was looking towards the far end of the field. The rider had been cantering the stallion back and forth but now he wheeled and came towards them. Elsie Manigault, ignoring or not hearing her son's reply, turned and called to the boy who had opened the gate. "Bob, ride over to Miss Lewis' and tell her that Mrs. Chapman won't be back till after lunch. Take Sinbad. But no tricks!" She looked at her watch. "You be back here in half an hour or I'll have something to say to you."

"Yas'm," the boy said and slouched away.

The rider was within a few hundred feet of them. Suddenly, as if a voice had spoken from a judge's stand, he put the horse into the "Allen walk." He rode, Catherine noticed, with the ease of a professional, holding his right rein so tight that the horse's steps curved inwards, as if he were playing for place at an imaginary rail. He was getting more out of the horse than she had been able to get. Indeed, he was forcing him to the point where if he had not been well trained he would have broken.

He dismounted and, touching his cap, handed her the reins. "Much obliged, Miss," he said.

"Mrs. Chapman, this is our trainer, Joe Marble," Mrs. Manigault said.

Tom Manigault laughed. "He ain't my trainer. I raise Herefords." He pronounced the word "Herford" farmer fashion.

Marble, grinning briefly, touched his cap to Catherine again and, moving off a little way, stood looking at the horse. Elsie Manigault, her head thrown a little back, was looking at him, too, through half closed eyes. "I think he's too shallow in the flank," she said abruptly. "Joe, don't you think he's too shallow in the flank?"

"He ain't filled out yet," Joe Marble said.

From the stable a great iron bell began to ring. Marble moved over to where Catherine stood. "I'll take your horse, Miss," he said and, slipping his arm through the reins, started for the gate. Two or three of the negro hands followed him. The rest, still sitting on the ground, reached into the bushes behind them and dragged out tin buckets from which they took sandwiches of cold meat and bread.

"Tom," Mrs. Manigault said, "are you coming to lunch?"

He nodded slowly and, going over to the tractor, unscrewed a cap, then slowly screwed it on again.

"You have to change, you know," she called as she walked away.

The bell was still ringing. Miles away another bell began to toll and then another, until the air was full of the clangor. Catherine, moving beside Mrs. Manigault over the brown stubble, listened to the sounds that seemed to come from a forgotten life. A day like this in early fall; the long morning nearly at an end; the green all unstirring. You would pause under a tree in an interval of one of the lonely games you played in that old yard and suddenly the first bell would ring out and then, as if it had been waiting for the call, another would answer and still, while you stood tranced in the reverberating air, another: Swan Quarter, Big Pond, the

Beekman place, all those fields, those woods had voices and spoke through the bright, hot air to each other.

"It is half past eleven," she said absently. "The farm bells always ring at half past eleven."

Mrs. Manigault laughed. "And we lunch at twelve . . . We keep all the customs of the country."

She opened the gate. They passed through into the lane. A stone's throw away Joe Marble was going up the steps of a neat, clapboarded cottage. Farther down the lane were other cottages, each enclosed by a neat picket fence. There had been tenant houses along here in Cousin Robert's day but they had been tumble down, unpainted affairs or ancient, weathered log cabins.

It's time I said something. "You've made a great many improvements, Cousin Elsie."

Mrs. Manigault had turned and was looking down the lane. "I've made some costly mistakes," she said in her rich, humorous voice. "Keeping those cottages painted and repaired costs me as much as keeping up a house in town."

Her eyes were the same startling blue as Tom's. Her white hair curled all over her head in short, crisp curls. With her leanness and her lithe gait she could have passed for a girl, but for that white hair.

". . . the *screens,*" she was saying. "You have no idea of the things they will do to screens."

"Why do you live here, Cousin Elsie?" *I ought not to say that.* "I mean— You used to have such a charming place on Long Island."

The woman's eyes had been hard when she spoke of the tenants but now her expression changed. She looked uncertain, even dazed, as if she did not know how she came to be walking along this lane, among these houses. She spoke slowly:

"Robert Manigault died in 1936. In March. Tom was in Guatemala. I had to come down by myself. After the funeral . . ." her words trailed, came to a stop. She fixed Catherine's face with the bright, blank eyes of one who is trying to remember, then began speaking again, her voice firmer. "After the funeral I was here by myself for a few days. It rained and rained. If you could have seen the house! That room he died in. Ten days after he was dead you could have cut the tobacco smoke with a knife. And other odors. That old woman who did for him was filthy. I had to get women out from Carthage to clean." She laughed suddenly. "And then I had to stay over a day. To sign some papers. A man drove up. He asked if I didn't want to repair the house. I went outside and looked at it and I looked at him and I said, 'It's no good repairing it. The thing is to tear it down.' He was Irish. He said, 'I'll help with that, too, Lady.' A little contractor from Carthage. He did quite well on the woodwork. . . . Naturally I couldn't leave the site bare . . ."

They left the lane and set foot upon the turf. More massive mounds of box and above them the house, gleaming white through the leaves, beautiful but in this air, among these trees, unreal. *A crazy house. A fairy house. She waved her wand and it sprang up, just as she waved her wand and the old walls fell down. She ought not to have torn down those old red walls . . .* "It is a beautiful house, Cousin Elsie."

"Roy says he is satisfied with it. I think he's one of the best architects in New York, don't you?"

"I never saw any of his houses except the Cravens' place on Cape Cod. I liked that."

They crossed a wide, flag-stoned terrace and entered the hall. Mrs. Manigault hesitated, then: "You can see the house after lunch," she said and led the way upstairs.

The room they entered was evidently her own bedroom. An enormous room, deeply carpeted in white. Its expanse of yellow wall was broken only by a huge oval mirror and curtains of ice blue satin.

Mrs. Manigault had hesitated again on the threshold. Catherine realized that she was expected to make some comment. She said, "What a lovely room . . . Is that mirror Regency?"

Mrs. Manigault sank down on a chaise longue drawn up to the window. "Second Empire. You remember all those French things at Mount Vernon? And the bedroom where Martha is supposed to have spent her last days. Roy says the interior decorations overcame her. The poor thing wouldn't have shut herself up like that if the house hadn't been furnished so grimly. Lafayette ought to have sent some things over for her instead of all that brass work for the parlor. Roy wanted me to have Venetian red walls in here but I told him I must have yellow . . ." She waved a hand towards the bathroom. "Go on, dear. I'll rest here a minute."

Catherine, rouging her lips in the mirror over the washstand, could see her hostess' figure extended full length on the chaise longue. Mrs. Manigault lay on her back on the apricot-colored damask, her hands flung down at her sides.

Her head was tilted back. Her eyes were closed. Without the brilliant eye the face looked pale, even a little pinched. Fine wrinkles radiated out from the eyes. Deep lines, curving down from the nostrils, enclosed the mouth.

She was showing her years. Was she getting mellower as she grew older? Her greeting a moment ago had been almost effusive and here they were, washing up together, almost like two girls. Catherine recalled visits to the Manigault place at Sands Point. Everything had been very formal. Separate bedrooms for married couples. Footmen. Two newspapers sent to your room in the morning. Something old-fashioned about the Manigaults. They had never kidded their money the way it was *chic* to do nowadays. Elsie, indeed, had been inclined to show off. Jim had had a run-in with her once over what he called her insufferable pretensions: one Sunday evening when they had been debating whether to go home that night or drive in early Monday morning and the butler had come and said that Mrs. Manigault would not be down again. Poor old Martin quite embarrassed and Jim, furious and very Middle Western: "Her Ladyship is indisposed? Well, tell her that the Duke of Sixty-fifth Street got sick at his stomach and had to go home . . ." It was shortly after that that Edward Manigault died and his widow went abroad to live . . .

They went downstairs. Lunch was served on the terrace. No damask cloth, no footmen. Instead a young negro man, stepping quickly about a glass-topped, wrought-iron table to set iced melons before them on crystal plates.

"Sit down," Mrs. Manigault said. "We never wait for Tom."

Roy Miller sat beside Catherine. He did not speak until he had finished his melon, then he pointed to the rectangle of greensward that, bordered on three sides by tall Lombardy poplars, stretched out from the terrace. "Do you recognize that?" he asked her. "I got the idea from your ancestral home."

"Swan Quarter?" she said in surprise.

"No. Castle Hill, in Virginia. Have you never been there? I thought all the Lewises and the Riveses were related."

"I believe they are. I descend from another branch of Lewises, not so grand. 'Irish John who slew the lord' was the first one in this country." She looked off over the smooth turf. At the far end a vista in the tall trees gave on the field they had quitted a half hour ago. "It's very effective," she said. "Is it like this at Castle Hill?"

"Rather. Only the Princess Troubetzkoy has tree box two hundred years old. Elsie has to do with Lombardy poplars."

"They do very nicely," Mrs. Manigault said.

Miller smiled at Catherine. "What had the lord done to enrage Irish John?"

"I have no idea. Some dispute about rent, I suppose."

He laughed. "Ah, the South! Now my father came to Minnesota from Bavaria in 1897. After a dispute about rent. But there's no romance in that . . . Here comes Tom. Tom, I planned to rise with the dawn and ride with you over the dewy fields but I had a nightmare that left me exhausted." He turned to Catherine, arching his black brows. "My father's boots on the dairy floor."

"Wooden?" she asked.

"Wooden, re-enforced with copper. The floor was concrete . . . The milk train left Bauer's Crossing at four o'clock. We used to start milking the cows at three."

"That was forty years ago, Roy," Tom said. "Haven't you caught up on sleep yet?"

He had had a bath and had changed into a white linen suit. Against his deeply burned skin his hair and moustache looked almost white. It was not the tan that one gets on a golf course but the red burn that comes from exposure to wind as well as to sun.

He realized that she was looking at him and he looked up, fixing her face with the same intent gaze that he had given her in the field. There were yellow flecks in the iris of his eyes. That was what gave the eye that peculiar, almost golden sparkle. His mother's eyes had the same golden flecks, but except for the eyes mother and son were not much alike.

He had protracted his gaze so long that she felt that she must take some notice of it. She recalled that he was ten years younger than she and that the last time she had seen him he had been an awkward boy of fifteen. She dropped her eyes to her plate, then, raising them, fixed him with a level look and leaning a little forward spoke in the indifferently aimiable tone that in that part of the country and particularly in her own family connection was used in addressing a younger cousin.

"Tom, how long have you been farming?"

"Three years," he said. He did not take his eyes from her face. His gaze, indeed, grew bolder.

She was annoyed to feel her color rising and turned to his mother. "Cousin Elsie, did you bring your cook down?

This—" she indicated the *coq au vin* that had just been served from an earthenware casserole—"This chicken seems mighty subtle for Montgomery county."

"I brought Bertha down but she wouldn't stay," Mrs. Manigault said. "She undertook to train a cook for me, though, before she left. A young colored girl. She learned very quickly." Her eyes strayed around the table and came to rest on her son's face. Her voice, to Southern ears, usually too incisive, too impelling, softened a little. "We haven't done so badly, as far as food goes. Have we, Tom?"

"It tastes just about like it does everywhere," he answered without looking up.

Mrs. Manigault laughed abruptly. "There's not much pleasure in catering to Tom. I believe he'd just as soon eat his lunch in the field, the way the men do."

"I ate with Ed Helfrey one noon," Tom said. "He had black-eyed peas. They tasted good."

"What kind of labor do you have, Tom?" Catherine asked.

"Half negro and half white," Mrs. Manigault said before he could answer. "It's the whites that make the trouble." Her face flushed. Her fine eyes sparkled. "I go east and hear my friends sympathizing with the Southern tenant farmer, and I wish they could see that row of cottages out there." She gestured with her fork. "Those six cottages are better equipped than any tenant houses in the whole country. Running water, refrigerators, electric stoves. Screens . . ." She gave a little gasp. "I thought when I equipped them that the people would have the decency, the common gratitude to take care of the things, and yesterday," she looked directly at her son, "yesterday I came by the Helfrey house and what should come out but the cat! They'd just finished cutting a hole in the screen so she could get in and out easily."

Roy Miller laughed and, calling the butler over, told him that he wanted more chicken.

Catherine had a mischievous impulse which she did not resist. "There used to be an arrangement like that at the old house," she said. "A tiny, swinging door that Cousin Robert's cat could push right through. As a child it fascinated me. I tried to get them to install a similar system at Swan Quarter but Mammy said it was a bachelor trick."

"He lived in great squalor . . ." Mrs. Manigault said absently. "I believe they are even worse about clapboards than they are about the screens. We have to keep a man going the rounds. If a board gets the least bit loose they'll rip it off and use it for kindling."

"You should install oil furnaces so they wouldn't be tempted," Roy Miller said.

There was the grating sound of a chair being pushed back. Tom Manigault was on his feet. He was staring at his mother, his face the color of old brick. He uttered a harsh, indeterminate sound, then bending his body awkwardly towards the others, left the terrace.

Roy Miller had been about to help himself to chicken but now he indicated with a gesture that the boy should take the casserole away. "I have no appetite," he said in a melancholy voice.

The sparkle had died from Mrs. Manigault's eyes but the painful color lingered in her cheeks. She picked up her napkin and made a fanning motion with it, then, laying it down again, emitted a short breath. "Really, Tom is impossible," she said. "Why he imagines that he can ever become a farmer! He can't even command the men, with a temper like that."

"I'm afraid I said something that annoyed him," Catherine ventured.

Mrs. Manigault shook her head. "No. It is what I said. About his uncle. That wretched man, who never did a stroke of work in his life, hardly even took a bath!"

"Elsie . . ." Roy said.

She turned to him. "Really, Roy, it's got to the point where I hardly dare open my mouth." She flung her hands wide. "And I am a person who is used to speaking her mind. I have always been active. I have even, as you know, been thought to have some organizing ability. And here I sit on a thousand acre farm and if I so much as mention damage to property, *my* property, my son flies into a rage and leaves the room."

Miller bent his black gaze on her. "And you can't go abroad to live now," he said thoughtfully. "You're one of those people who are cooped up in America."

A negro maid stood motionless in the doorway for a second, then came forward. "Mister Marble at the back do'," she said in a soft whisper, "say kin you come to the stable a minute?"

Mrs. Manigault rose swiftly. She stood for a second, her eyes downcast, her fingertips resting on the table edge and in that brief interval seemed to regain her composure. She came to Catherine, who had risen too, and took both of her hands in hers. "You must come to us again soon, my dear," she said. "We aren't always so quarrelsome," and bending she kissed her cheek. In the doorway she turned around. "Make Roy show you the house," she called. "I want you to see it."

Servants came and noiselessly cleared the table. Catherine crossed to the other end of the terrace and dropped down into a long garden chair. Miller sat down near her.

"Do you want to see the house?" he asked.

She shook her head, thinking that she would go as soon as she had smoked another cigarette, and watched the shadows move across the rectangle of turf. Squat, round shadows now, but by late afternoon they would be long lances covering the whole plot. It was a charming and unusual effect, however come by. She thought indifferently of the man at her side. He probably had a good eye for that sort of thing and picked it up wherever he went.

After a little he spoke, absently. "I suppose you'd be on Edward's side?"

"Edward?" she said in astonishment. "Cousin Edward Manigault? He's dead."

"The evil that men do lives after them . . . He never should have married that girl."

She turned on her soft, cushioned seat and stared at him. "Cousin Elsie?"

He nodded. "I've known Elsie for years. At eighteen she was glorious. Those eyes . . . and her hair was a wonderful auburn, rather like that horse you rode over here. She moved like a goddess—of course, still does . . . Manigault . . . I could see what she saw in him, all right. He was a handsome fellow. Brilliant lawyer and all that. And then there was the family angle. But he was a *gray* man. Gray even then, at twenty-eight. I told her, but she wouldn't listen. That was because she knew I was in love with her." He smiled. "You may not believe that but it's true."

She straightened up, and as she did so saw him turn his head with a quick, birdlike motion. He was a big man but all his movements had an almost feminine quickness. And yet in repose his flesh, particularly the dark, sagging cheeks, looked soft and flaccid. She had set eyes on him only two hours ago but he had already made her several confidences, either by word or by one of his meaningful glances, and now he was talking about matters that concerned neither him nor her.

She stood up. "You talk a good deal, Mr. Miller," she said coolly. "Half the time I don't know what you're talking about." She extended her hand. "Thank you for a pleasant morning."

He did not answer but, moving to the door, summoned a servant who went out and, coming back almost at once, reported that her horse was waiting. Miller walked out with her. Standing in the drive while she mounted he gave her another of the liquid, dark glances that she so much disliked.

"You must come back," he said.

V

Mrs. Manigault rose from her knees. She flecked imaginary dust from her skirt, then, bringing the palms of her long, finely shaped hands together, brushed them for a moment against each other. "She ought to be drenched again," she said. "I've thought all along that she ought to be drenched oftener."

Joe Marble looked down at the sick mare. "She's got that apomorphia in her now," he said. "Maybe we ought to let her lay."

"We ought to be doing something," she told him impatiently, and stepped out into the runway of the stable where Tom leaned against the wall, whittling. "Do you think we ought to have Dr. Kelly out again?" she asked.

"If you want to," he said.

She moved closer to him. "What do you *think*?" she asked in a low, urgent voice.

He held the whittled stick out in front of him, examining it through half closed eyes as if he intended it for some special use. "I think he's just barely got back to town," he said. "And I think there are other sick horses in the county and maybe he wants to spend his time going to see them instead of running out here every half hour."

Joe Marble closed the door of the stall and came out and stood near Tom. He put his hands in his pockets and, for a moment, occupied himself with alternately balancing on his toes and sinking with a little click back on his heels. Then he spoke: "There's one thing. Dr. Kelly ain't going to come out here twice in one day. To drench no horse, that is. Or cow, either."

Mrs. Manigault's finely cut nostrils quivered. "He'll come if he's paid to come," she said.

Marble was silent, continuing his rhythmic balancing. From outside the stable door where some negro boys were lounging came a sudden burst of laughter. "Go 'way from here, boy," a voice said, "I know where you was Sat'day night."

"Ananias," Mrs. Manigault said without turning around.

There was the sound of a scuffle and quick steps on the sawdust. A negro boy appeared, carrying a saddle under his arm.

"Haven't you finished those saddles yet?" she asked.

"Naw'm, but I'm working on them," he said, and stood, surveying the group with dark, curious eyes. When she did not speak again he turned and carried the saddle outdoors.

Mrs. Manigault looked at her watch. "I'm due in Carthage now," she said. She turned to Marble. "Joe, I'm going to telephone you from town. If the apomorphia doesn't take effect by the time I get in there I'm going to call Dr. Kelly. He can either come out or lose his practice here."

"Yes'm," Marble said quietly.

She walked away. The young negroes squatting in the doorway got up quickly to let her pass. As Ananias sank down again he turned his head and exchanged a slow smile with another stable boy.

Tom Manigault flung his stick from him, but he seemed with a start to recognize that Marble was still there. "I'm going over to the pond," he said abruptly. "You want to come along?"

Marble shook his head. "I better stay with the mare."

"Suit yourself," Tom said and started for the house. His mother was still in sight, walking ahead of him up the lane. From this distance she looked almost like a girl, swinging along in her brown skirt and tawny sweater, an old hunting cap crushed down on her hair. In a few minutes she would be in her car and driving to town. She drove well for a woman, but she drove too fast. She did everything too fast. The negroes were on to that. "Mrs. Make Haste" they called her. He had heard the phrase on one of the hands' lips. It had been accompanied by just such a slow, significant smile as Ananias had turned on the other stable boy a moment ago.

He went up the back stairs to his room and, getting his gun and slipping some shells into his pocket, set off over the fields. As he went he glanced up at the sun. At least two hours more of daylight. But it had been hot and dry for over a week. The doves might be coming in to water early.

He went as the crow flies, climbing fences when he needed to, veering to the east until he came to the woods. There was a path but it was not much travelled. He doubted if anybody had been in here since dove season last year. These woods had not changed much since he was a boy and used to come in here at night with his uncle, hunting coons. Over on the south side of the woods there were a lot of wild grapes. They used to go there first and then if the dogs didn't tree they would head for a persimmon thicket on the other side of the pond. Caleb Jones always went with them. He had the best coon dog on the place, Old Ride. The niggers didn't seem to hunt coons as much as they used to. At least he never heard them talk about it. And if there was a good coon dog on the place he'd never seen him. But in Uncle Robert's day, that fall that they'd let him stay down here till Thanksgiving, every nigger on the place had a dog that would tree.

He came out on the edge of the woods. Before him a field of millet was pale green in the late sun. Beyond the field the pond lay, fringed with willows. He sat down on a stump, not taking any pains to conceal himself but sitting there carelessly, his gun across his knees.

He was thinking of a man who was dead. The uncle with whom he used to hunt these woods at night. The old man's asthma was always worse in the fall. Sometimes, on those cold nights, moving against slapping branches, through the roofless cave that Caleb's lantern, swinging its wet gleam on the leaves, drew always behind it through the woods, you would think you were alone, left behind, and then you would hear the wheezing, and the tall shape would be beside you and then moving away to some black trunk. "Damn," he would say, *"Damn!"* and, bringing both arms up over his head, lean his forehead against a tree while his body writhed in a convulsion of coughing. It made Caleb mad. He would call out. "Lord, Mister Robert, you roar so loud you'll scare the game."

Afterwards, in the south bedroom that smelled of tobacco and coal smoke, you would sit sunk back in a big rocker while Uncle Robert stood at the washstand washing himself. When he came back to the fire he would have on his red woolen dressing gown, his thin shanks in their long, ribbed underwear showing beneath it. A shelf called a "footman," hooked over the grate, supported the kettle from which he had got his bath water. There would still be water in the kettle, and he would take the bottle and glasses down from the mantel and, holding them up, blow off the light layer of coal ash before he mixed the toddy: three or four glasses for himself, one light, full glass for the eleven-year-old boy.

That was when he would talk, not so much about coons as other game, and dogs that he had known. One old pointer that he called "Trecho" that could out-run anything in the country. One of the first things that he himself could remember was riding on that old fellow's back.

Two doves flew out of the wood, fluttering and dipping. He stood up and fired. One was falling, plummet-like; the other was half way to the pond. He fired the other barrel and missed it.

He walked out into the field and, picking up the dead bird, thrust it into the pocket of his hunting coat and went back to the edge of the woods. Four more birds flew out towards the pond. Standing erect beside a dead chestnut, his face upturned to the pale sky, holding his gun in readiness, he gave no sign that he had seen them.

When I was a kid I was always off at school or at camp, or sometimes they would let me stay down here. I never saw much of her. I never paid much attention. But she was running the show even then. The old man knew it, but he didn't give a damn.

The old man never tried to tell me anything. Uncle Robert did. That time I came in from shooting ducks and he was sitting there in that old rocking chair by the window and he said, "Tom, what are you going to do when you get out of college?"

I tried to kid it, said, "Uncle Robert, I like to box pretty well, but there's not another damn thing I like to do except hunt and fish, though I wouldn't mind having a boat."

He shook his head, said, "Don't fool, Boy. Your father was lucky. He had his law—and I've got my asthma. You better get hold of something."

I'd just been out to Wyoming, with Bill Jackson, so I said I'd make them buy me a ranch and raise cattle.

He said *"Ranch?"* like he was kind of mad and then he laughed. "You're a natural man," he said. "You better get on the land. Then the seasons'll take care of you."

An old man can smell of stale tobacco and not change his clothes often enough and still have more on the ball than anybody you ever knew. I don't know what it was. And now he's gone. But a minute ago, there on the path, it was like he'd just been along here, like I could catch him if I'd hurry. I've had that same feeling out in the field. Never in the house. He couldn't come back to that house. But the dirt's still here. Best Hagerstown clay, going straight down to China. She had to leave it. She couldn't pay enough to have that carted off.

I came up the drive and the old house was gone and there wasn't anything in its place, only the heap of old red bricks that you could tell hadn't fallen down but had been torn down by somebody who hated the house, had hated it all these years and had waited till there was nobody by and then had struck. I thought: I will kill the one who did this. And then she came out from behind the well house and said "Tom" and I had to stand there and let her kiss me.

She wanted me off the place while they buried the old house under the new one. She showed me a letter she had from the Archaeology Department, from old Percy Rountree, saying I could go on the South American dig. I said, "How much did you offer him? I know Perce Rountree. I took his course. He knows how dumb I am. He wouldn't let me polish a bone, unless you paid for the privilege."

We were living in the Carthage hotel that winter. That old dining room always smelled of onions. She got that look about her mouth, said, "You did well in history. I used to think you'd be an archaeologist . . . I'd like you to be something really distinguished, Tom." I said, "You'll see what I'm going to be," and pushed my chair back and got out.

I didn't have but ten dollars on me. That old trout at Sparks and Johnson didn't want to let me have the wire. Didn't know who I was. "I'm Tom Manigault," I said. "I want to run a line of fence straight down the middle of our place. My mother and I are dividing the farm. We don't get along."

After we divided the farm she might have gone away, the way she does when she gets tired of anything, but she found out that the bank was letting me have the money to buy the machinery and the Herefords too, and she wouldn't leave. That was what started her on the horses. She doesn't know horses. She let the best colt ever was on the place slip through her fingers—that stallion that girl rode over here today. She's tired of the whole business. But she won't give up. It's the first time she's ever been crossed in her life. She can't understand it.

What in hell was the old man doing with her all those years? Why didn't he beat some sense into her? Was it because she had ail the money? They tell me the old man was a brilliant lawyer, but he never had much to say. It was like they'd sucked it all out of him. He never even seemed to take any comfort in liquor. I saw him drunk once: that night I came down to the library to ask if I could go to the football game and he and Judge Carter were drinking and talking history. He had his hand on the judge's knee. His eyes looked soft, like he was going to cry. He was sort of whispering: "I'm a damned carpetbagger. Only I went North," and then he saw me and his face tightened up, and he said, "What do you want, Tom? You ought to knock before you enter a room."

He lowered his gun slowly. He glared at the evening sky which was now full of the circling, dipping birds. He said in a whisper: "You ought to have left me the land. It was mine. It was *ours*. Before she was born."

Two birds had lagged behind in the flight. He raised his gun and fired. One flew on. The other was falling, not plunging straight down but fluttering a little. He walked out into the field. He had to tramp some of the millet down before he found the wounded bird, crouching on the ground, among the green stalks, as if on a perch. There was blood in its eyes. A drop of blood was hanging on its beak. It gazed straight before it out of its bloody eyes. He could see the white feathers on the throat puff out with its slow breathing. He tramped more of the millet down, backing off to shoot it, then picked it up. For a second he wanted to lay it back on the ground. It seemed different from any other bird he had ever shot. And then his hand carried it on down into his pocket, to lie beside the other dove.

VI

THE DARK CARPET was ornamented at intervals with arabesques of a lighter shade of gray. In the morning the figures were scarcely visible. But by four o'clock the sun had climbed over the top of the building. Its oblique rays, broken by intervening boughs, spattered the patch of carpet near the window. The figures—stars, bars, crescents—swam in quivering light, then composed themselves into faces. There was one that, surging farther forward than the rest, seemed about to break from the imprisoning wave.

The man at the desk stared at the grinning face and got up and pulled the cord that lowered the blind. With his foot he straightened the roll in the rug and stood with his hand on the cord, listening. A few minutes ago there had been sounds outside his door, the steady swish of a mop over marble, an occasional footfall. But for some time now there had been no sound, either in the corridor outside or in the room. He glanced at the floor. The faces all lay in a row, staring at the ceiling. One seemed to gaze more sharply than its fellows. The man's black brows drew together. Crow's-feet showed suddenly about his deep-set eyes. He jerked the cord so that the blind dropped even lower and went back to his desk.

It is not the sun coming in at the window or the way the figures make themselves into faces. It is the silence. It does not come out until the last student has left and then it gathers in the great hall and creeps up the stairs. It takes a while, but it is punctual. Every day at four o'clock the silence comes.

He closed the book he had been reading, slipped it into a brief case, took his hat and stick and went down the wide stairs. On the pavement outside he hesitated, looking down the street, a big man, a little stoop-shouldered, his black hair already touched with gray at the temples. He settled a battered Panama hat on his head and walked towards the river. Fallen leaves from the gingko trees that grew in a row before the building rustled under his feet. One yellow, fan-shaped leaf clung to the toe of his shoe, quivering a little each time he set his foot down. Behind him on the street a voice was calling. His shoulders hunched forward, he walked a little faster. The voice called again:

"Chapman! . . . Hey, Chapman, wait a minute!"

Chapman slowly turned around. A young man in a brown suit was running towards him. His brief case slapped his stout legs as he ran. He came to a stop under a gingko tree, and, pushing his hat back on his auburn curls, mopped his broad forehead.

"It's hot," he said. "You going downtown? Want to split a taxi?"

Another fan-shaped leaf fell from the tree. Chapman watched it drift past the brown, woolen shoulder. Fifteen years he had trodden these yellow leaves under foot. It seemed to him that they were falling earlier than usual this year. But then it had been very dry.

"I'm not going downtown now, Wilkins," he said.

Wilkins glanced at the building on their right. "I'll go over to the library," he said. "You give me a ring when you're ready."

Chapman looked into the eyes that were the same shade as the wool that clothed the athletic body. Wilkins' wife chose his suits for him, with the purpose, she had once told Chapman, of enhancing the rufousness that she and other women found so attractive. It made him look like a cocker spaniel. But perhaps that was what she and the other women wanted.

"I'm not going downtown now, Wilkins," he said again and turned and walked a few steps and paused. "Hell," he said and looked back over his shoulder, "I haven't had any lunch."

Wilkins was already hurrying towards the subway, but, hearing Chapman's voice, he, too, paused. "Sure," he called, "I didn't mean to hold you up, Jim. I just thought we might split a cab."

"I'm glad he didn't wag his tail," Chapman said under his breath and turned into a drugstore that carried in its window a sign: "Hot Chocolate—Lunch—Soda."

The place seemed dark. There were half a dozen tables, all occupied by students. As he sat down on a stool at the counter a voice beside him spoke respectfully: "How do you do, Sir?"

"Hello, how are you?" Chapman said, without turning his head, and gave his order to a waitress.

He took a bite from the sandwich she set before him, then pushed it aside. He drank lukewarm coffee. The place, which had been in a hubbub when he entered, was now quiet, except for an occasional giggle from some girl. The students were not used to seeing a professor sitting at this counter. He ought to have known that this was a student hangout. What had possessed him to enter it?

He drained the last drop of coffee and was about to slide from the stool when the boy sitting beside him spoke again: "I'm on the eighth canto now."

Chapman turned around. Brown eyes darted a quick look

at him from behind horn-rimmed spectacles. A wide mouth trembled into a smile. "I'm on the eighth canto," the boy said again.

The big man in the Panama hat stared, unable at first to identify the student. Then he remembered that several weeks ago Dick Reynolds, of the History Department, had brought a boy to his office: Paul Bergmann. Paul was carrying five subjects, Dick said, and reading the *Inferno* on the side and doing very well with it. It had been ten years since Reynolds had read any Italian. Some of the boy's questions had stumped him. Would Chapman help them out? The three of them had read the first two cantos, sitting there in the office. Chapman had not seen the boy since then, until today.

He smiled, "Ah," he said, "you're in the Fifth Circle."

The boy nodded. He drew a small volume from his pocket and, leaning over the counter, read eagerly:

> *Mentre noi correvam la morta gora,*
> *dinanzi mi si fece un pien di fango,*
> *e disse: 'chi se tu, che vieni anzi ora?'*

He looked up. "Why did Dante dislike Filippo Argenti so?" he asked.

"I don't know," Chapman said absently. "Probably because he was a member of the Adimari family. Dante had a great many enemies, you know . . . That's very fine. And it's very fine later on, about the Fallen Angels."

The boy's eyes fixed Chapman's face intently. "I'd like to hear you read the fifteenth canto," he said, flushing. "Professor Reynolds said he heard you read it once and he's never forgotten it."

Chapman's eyes were on the book, lying open between them at the first page of the poem:

In the middle of the journey of our life I came
to myself in a dark wood where the straight way was lost.

. . . . . .

I cannot rightly tell how I entered it, so full of sleep
was I about the moment that I lost the true way . . .

With an effort he took his eyes from the lines. He looked through the glass door out on to the street. Above a half-built brick wall a white cloud floated in the formless blue. He looked down at the boy, suddenly set his hand on his shoulder. "I'll be glad to," he said. "Come around to my office some time. We'll get Reynolds and have a reading."

He went past the silent tables out on to the street. In the restaurant behind him the noise that had died on his entrance had risen again, as suddenly as shrieks from Dante's damned . . . *In the middle of the journey of our life I came to myself in a dark wood* . . . A cab was cruising up the street. He raised his stick and went towards it.

"Fifty-two East Seventieth Street," he told the driver.

The cab turned into Riverside Drive. Chapman sat looking out over the water. There was little traffic: a tug, moving slowly upstream, a destroyer riding at anchor, and farther down the great liners tied up at the docks.

The driver gestured towards the *Normandie.*

"They say that thing cost eighty million dollars. I bet them Frenchmen wish they had their money back."

Chapman roused himself. "What's the news?" he asked, "I haven't seen an afternoon paper."

"They're bombing hell out of Buckingham Palace," the driver said. "That's where the King lives, ain't it?"

"That's where the King lives," Chapman said and looked back up the river.

The scarlet and white hull of the *Normandie* framed the gray, receding bulk of the destroyer. I wonder how many we've got? he thought. His eyes returned to the giant liner. Her paint was getting dull. She had been lying there for two years. This is September, 1940, he thought. France has fallen. The Germans are dropping bombs on Buckingham Palace. His gaze lifted to the horizon where, past the Palisades, sparkling, open water showed a few white-caps. Italy . . . he thought. God help her! I may never see Italy again. I wish I had gone over last summer. I would have gone if I had known how things were going to turn out . . . But I did not know how things were going to turn out, *so full of sleep was I . . . about the moment that I lost the true way . . .*

They had left the ramp. The cab was stopping before an apartment house on East Seventieth Street. Chapman, leaning forward, his hand in his pocket, looked up at a window on the sixth floor where potted geraniums grew in a row. He withdrew his hand from his pocket, sank back against the leather cushion.

"I told you Sixty-five East Sixtieth," he said.

The driver shrugged his shoulders, turned down Fifth Avenue and then east on Sixtieth Street and stopped at Sixty-five. Chapman paid him, got out and took the elevator to the fourth floor.

He let himself in, set his stick in a corner and laid his hat on the table. The hat was soiled. There was a break in the straw of the crown. Each day when he came in he noticed that break and told himself that it was time to get out a felt hat, but every morning when he went out he forgot and put the straw hat on again. He decided that if he put the hat out of reach he might find it easier to discard it and, stooping, laid it on the floor under the table, then went into the living room. At this hour of the day it was flooded with western light. There was no one there, but from behind a closed door came a soft crooning. Lily, the colored maid, was still at work in the kitchen.

The door opened. Lily appeared.

"Mister Jim, you going to be here for dinner?" she asked.

"No," he said, "I'm going out."

Lily went into the kitchen and came back, wearing her hat and coat. She stood, drawing on her gloves, *suede* gloves of a light shade of gray that he thought his wife must have given her. "I be here in the morning," she said. She went towards the outer door and paused. "The orange juice is all squeeezed," she murmured. The door closed behind her.

Now that she was gone the place seemed very quiet. *I be here in the morning . . . The orange juice is all squeezed . . .* Standing in the middle of the floor, rolling a cigarette between his fingers, he frowned. The words had some sort of ritual significance that for a second eluded him. Then he remembered. His wife, Catherine, maintained that she could always tell when Lily was getting ready to take a day off. If, after assuring you that she would be here bright and early in the morning, she told you that the orange juice was already squeezed for tomorrow's breakfast, you knew that she would take tomorrow off. If she closed the door and then put her head in again and said, "Honey, take care of yourself," you knew that you would not see her for at least a week.

Catherine—a scatter-brained girl; she had once lost a good fur coat by misplacing the address of the irresponsible furrier to whom she had sent it for repair—was full of wise saws and rules of thumb for the management of negroes. She came originally from Tennessee, Lily from Arkansas. It seemed to him that they never communicated with each other directly but always in some code that apparently each of them understood. Catherine's first rule was never to advance a negro money on his wages and never to lend him money. Give him the shirt off your back—she would do it, too; one winter he had paid out two hundred dollars for radium treatments for Lily's old mother, who had died, anyhow, of cancer—give him the shirt off your own or anybody else's back but never do him the simple human kindness of lending him money—or you would lose your nigger. "My nigger," she called Lily when she was talking with her friends over the telephone, though Lily, to her face, was always, of course, "a colored person." Catherine had a patronizing, proprietary air towards negroes that he still found irritating after fifteen years of married life. Once they were walking on Fifth Avenue and passed a handsome, six-foot Senegalese soldier. "Heavens, I'd like to have him!" Catherine had exclaimed. They were just married. He did not know Southerners then as well as he had come to know them. He had thought for a second that she meant that she would like the negro as a lover, but what she meant, of course, was that she would like to have the fellow, a lieutenant by his bars, polish the floors of the apartment they had just moved into.

He had been about to turn back into the little room at the rear of the apartment that served as his study, but he flung himself down now in a chair that was drawn up before the empty hearth. Leaning back, the ash from his cigarette fallen in a long streak across his shirt front, he contemplated his wife's portrait hanging over the mantel.

It had been painted five years ago, when she was twenty-nine, or was it thirty? A fair, long-limbed girl in a pale dress, sitting on a marble bench in the midst of improbable foliage, absently fingering the ear of a white unicorn. Dave Koenig, who got his golds and his blues from armorial bearings and whose dream pictures full of heraldic animals were not so famous then as they are now, had insisted that Catherine be attended by a unicorn. Catherine, who liked to think that she was descended from the Monckton family, had been delighted. They had spent a great deal of time choosing the position the beast must assume. Koenig had finally presented it crouched beside the woman, or as he put it, *lodged,* with a *gorge* of roses and with a rose instead of the coronet *or* that showed the family's Crusading exploits, dropping from its pink mouth.

Both woman and beast had long, light brown eyes that seemed to follow you about the room. The painter, a loquacious fellow, always ready to talk about himself, had explained that you had only to have the sitter gaze at you continually as you worked to get that effect.

It had been one of those long time commissions. Koenig, who desperately needed work—it was the winter before the

Routh galleries took him on—had said that they might take as long as they wanted to pay for it. It had taken Chapman three years to get the thing paid for. Well, Koenig had taken his time painting it. A whole month he had spent on the eyes alone. He was astonishingly literate, for a painter, and had made jokes about his slowness, quoting Andrew Marvell: *A hundred years should go to praise . . . thine eyes . . . and on thy forehead gaze . . . Two hundred to adore each breast . . . But thirty thousand to the rest . . .* He had thought that the fellow had his nerve. Still, he had not betrayed his annoyance, until the end. "I suppose," he told them, "that eternity will finally seal your bliss with an individual kiss."

Koenig had laughed his head off. They had all three gone out to a night club where he had spent forty dollars that he could ill afford, buying champagne to celebrate the picture's completion. Why did women always want champagne? He had always disliked the stuff, even when he was so poor that he never got it except as a hand-out from the rich.

He gazed at the slim horn that sprang from the unicorn's forehead, fawn-colored, delicately ringed with green, as tender as one of the fruits with which Crevelli's Holy Infants toy. Koenig had been particularly pleased with it, as good a bit of painting as he'd ever done, he said . . . The champagne, he thought, might have been drunk, not so much to celebrate the completion of the picture as the eruption of that horn—on his own brow as well as the unicorn's.

He threw his cigarette from him. It fell short of the hearth and smouldered on the rug. He got up and stamped it out. He did not think that Catherine had ever gone to bed with the fellow. She was too fastidious for that. If she had had an affair with him she would have gone about it in a different way, made more out of it. A cold woman, really, moody or subject to sudden, inexplicable withdrawals.

She had been gone now for several weeks. The note she had written him when she left was still in there on his desk. He had not had any warning that she was going, and he had not had any communication from her except that note that she had left on his typewriter. The typewriter nowadays, not the pin-cushion. Short. He had it by heart:

"This evidently belongs to you. ["This" referred to the note that had caused all the trouble. From Edith, and, of course, indiscreet. Like a fool he had dropped it on the rug and Catherine had found it.] I have taken Heros with me. Please keep Lily. Your heavy coat is at the cleaner's. I have taken the car, Sinc'y, Catherine."

His big right hand came up and clenched at his side. *Sinc'y!* What a barbarous abbreviation, and she must needs use it in what was presumably her last communication with her husband!

The telephone rang. It was probably Edith. He had said that he would come to her at half past four. It must be five o'clock by now. The telephone rang again. From the wall, painted eyes watched him cross the room and sit down at the telephone table. He lifted the receiver and, without speaking, softly put it back in its cradle and held the cradle down. The buzzing finally ceased. His hand still pressing the cold metal, he sat looking across the room into long, light brown eyes that, fringed with soft, dark lashes would have been as much at home in a doe's shallow skull as in a woman's. He had never seen another human eye that so moved him.

He got up and stood in front of the picture. He said aloud: "Why did you do it?"

When there was no answer he walked to the window and stood looking down at the street. Between the gaps in the roof-tops he could see a stunted, yellowing vine swaying on a miniature lattice. He knew the roof it grew on. Catherine and her mother had had an apartment in that house when he first knew them . . . It was to that apartment he had gone when he was first released from the task that for so long had occupied all his energies.

For four years, while he was writing his History, he had lived on forty dollars a month, in a furnished room over on Eighth Avenue. He had not cared what the place was like while he was working, and then one afternoon in April—it was soon after he had delivered the manuscript to Smith and Soldan—sitting at the same desk where he had worked for four years, he found himself unable to work, even to read, the smell of fish came in so strongly through the open window. He had, or thought he had, another book in him at the time. But it was too soon to start it. He packed a bag and went over to stay with Ed Ware and Bob Upchurch in a three-room apartment they had on Thirty-eighth Street. They were both working on the *Sun* then. He got there on their night off; they took him along with them, to "Agnes'."

Agnes Sewell, Catherine's mother, was then about forty years old. She had got a job and stayed on in New York after her husband died. They had an apartment over on the East River. The boys who worked on the paper with her had the run of it. There were two small bedrooms and a kitchen and one big room that looked over the East River.

They ate at a long refectory table at one end of that room. Mrs. Sewell cooked the dinner: an enormous Lobster Newburg in a casserole. She had waited till they got there to start it; she could not bear to split the lobsters.

There was an old-fashioned portrait on the wall, a primitive: a little girl in varnished boots, frilled pantalettes showing beneath her blue muslin dress, scattering grain to some game chickens. The oval face was dead-pan, the blue eyes expressionless. The little outthrust fist looked more like a fish than a human hand, but the frilling of the pantalettes, the surface of the muslin were executed with some finish, the rooster's breast showed a fine scarlet and his tail feathers a blue-green sheen that one would have thought it beyond the painter's powers to render.

Mrs. Sewell left the people she had been talking to and came and stood beside him while he looked at it. "My great-aunt," she said. "Her name was Amanda Morrison. The chickens are nice, aren't they?"

He told her that he found them delightful.

Bob came over to them, carrying a hatchet. "Is this the best thing you've got, Agnes?"

She looked flustered and helpless. He was to find later that one of her chief, if contradictory, charms was the constant deprecation of her own abilities. "Maybe the janitor . . ." she said.

He took the hatchet away from Bob. "You'll pound them to bits. Haven't you got a heavy kitchen knife?"

"I've got a butcher knife," she said.

"A butcher's knife ought to be fine," Bob said. "Agnes, I didn't know you slaughtered animals up here!"

The three of them started back to the kitchen, but Bob stopped to talk to somebody on the way. He and she arrived in the kitchen alone. She turned to him as soon as the door had closed behind them. "Would you just latch that?"

He wondered as he turned the key why she used so obsolete a word as "latch." The door was fitted with an ordinary Yale lock.

He stood in the middle of the room, out of the way, he hoped, while she found the knife. He had just that week got around to reading a book by Richards and Ogden, and as he stood there it amused him to follow a train of speculation that the idea of Basic English had started in his mind. Basic English, or for that matter, basic French or Italian or Portuguese was already in use, it seemed to him. The New Yorkers whom he knew used a "basic" whose range was determined by their geographical location in the city, the nature of their occupation, their financial condition. This woman who had lived in New York for fifteen years, and worked on a newspaper for her living, had just startled him by using a word un-basic to her vocabulary.

Holding up a heavy, ordinary kitchen knife she beckoned him to the sink where three large lobsters lay, waving their tails and claws. She stood with her back turned, her hands clasping her temples, while he cut off the tails and claws according to her directions. When he announced that he had finished she thrust a salt cellar into his hand. "Would you just salt and pepper them while I melt the butter?"

When the butter was melting in a saucepan she looked up and laughed. "I always say 'butcher knife' for 'kitchen knife.' It's because I was raised in the country, where they butcher things."

He laughed too, delighted that she should continue his own train of thought. She looked towards the door. "I thought I'd just lock them in the house while we got started," she said and laughed again. "That's another thing I say. 'In the house.' "

He could not understand why she should thus differentiate one of the four rooms of the apartment until she explained that, in the South, in the old days, the kitchen was always placed a little distance from the rest of the house. "And so I still say 'in the house,' " she concluded, "in this apartment, where you couldn't swing a cat by the tail!"

She had neglected to read the recipe before she started, and as she moved about the kitchen murmured formulae that she snatched in rapid readings from the cookbook, placed high on a shelf above the hurly-burly. She wore a black dress, chiffon or something, with long, flowing sleeves, tucked up so inadequately that they kept coming down and fluttering over the saucepan until he put out a hand and seized one of her round, white arms. She laughed and held it still. He knotted the soft stuff tight and felt the warmth of the kitchen all around him, and heard the laughter echoing from the other room, and it seemed impossible that he could ever go back to the room on Eighth Avenue where, each day, the mice that kept him awake at night left their droppings in the paper wrappers of the bread, the cheese, the coffee that would serve him for breakfast. . . . She was not a particularly pretty woman, not very intelligent, not even very witty, but he had never known anybody whose society so continually and innocently delighted him. Now she had been dead seven years. He had not even got to attend her funeral. There was a paper to be read at a meeting of the Historical Association; Catherine had had to go alone with the body down to that desolate old place in the country.

He had known Agnes Sewell for some weeks before he met her daughter. He had seen the girl for the first time one afternoon when he and Ed Ware dropped around to Agnes' apartment unannounced. The place was full of young people, three girls and some boys from Princeton. A pretty, dark-eyed little girl kept saying, "Please wait. She'll be back any minute," and then a tall, fair girl had said that they were all driving out to Princeton to swim and had suggested that he and Ed come along. Ed, who knew them all, said that he would be delighted, and he, not having anything else to do, had gone, too. There was a boy they called Stoney. It was his place. The family was away. They had rushed through the house, scattering a lot of servants out of the way, and down to the pool. It was on the edge of a wood; some sort of play cart was drawn up at one end of it, spilling over with flowers. At the other end flat, stone slabs made a row of monkish dressing rooms. He came out of his cell to find more boys come out from town, diving, yelling, splashing water on the girls. He dove, swam the length of the pool twice and sat down on the steps. The tall, fair girl who was Agnes Sewell's daughter, came, dripping, to sit beside him. She had on a white bathing suit. He had thought that she was a queer color—it was when tan first began to be fashionable. He himself, having lived indoors for five years, must have been the color of a fish's belly. The girl was the color of an under-done biscuit. She had long, fine legs that, leaning back against the step, she extended straight before her. The water ran over the round of the calf, swelled into drops and pattered on the stone as she lounged beside him, talking. The pretty, dark-eyed girl was Molly Eskers, her visitor, from Nashville. (Ed later married that girl but nobody thought about that on that day.) She herself had just finished school and had come to live in New York with her mother. "I couldn't have gone to Sweetbrier," she said, "if I hadn't got a scholarship, but I'm not really bright." Her mother had told her about him. She said he wrote books. "No," he said, "just one." He had said anything that came into his head, for while she talked—Catherine had a pleasant speaking voice, not high, like most Southern women's—he was experiencing a sensation he had never before experienced.

Ever since he could remember, at least ever since he had entered the Mount Hope high school at the age of thirteen, he had had the feeling of being driven. There was something he had to do before he died—like all sensitive adolescents he had brooded much on death. But he had known, even as a boy, that he must be prepared for the task that would absorb all his energies. He had read a great deal before he was eighteen. At nineteen he had entered the university here in the city. It had been a grind. He had paid part of his expenses by waiting table in the commons. He had had friends, Bob Upchurch and Ed Ware, for instance. He had never had the feeling of being left out. But there had never been enough time. And in his senior year he had got interested in the research that finally led him to write his *History of the Venetian Republic*. Albert Briggs, the venerable historian under whom he had taken his Master's, had encouraged him to undertake it. A month after graduation he was in Italy, working among Medici records . . . All the time since then had been like one long day, no, a long night, in which one hardly dares get up from one's chair or light a cigarette, knowing that the task must be finished before dawn shows at the windows.

But sitting there beside that girl, watching the water drop from the brown legs on to the dark stone, that feeling had gone away . . . It was a day in early summer. The place was all in shade, except for a square of sun-lit water at the far end of the pool where a handsome, red-headed boy kept darting about like a goldfish.

"Let's take a walk," he said to the girl, and they left the pool and went off through the woods.

It was a rich man's place. The undergrowth all cleared out and the paths barbered. It took them a while to get out of sight of the others. They came to a brook. Tall grass had been allowed to grow wild on its banks. There was a young willow tree. They sat down and watched snake-doctors skip over the water. He could not remember anything that they had said. They had not been there long when the tall grass behind them parted. The red-headed boy stood looking down at them.

He said, "Are you ready to go home?"

The girl said calmly, "No."

The boy came forward. "You'll go now or you won't go with me," he said.

"I'm not ready to go now," she said.

The boy went up to her and took hold of her wrist and said, "Come on, Kit."

He found himself standing beside them, his hand on the boy's wrist. "You heard what she said?" he asked.

The boy whirled around. He said: "Shut up. Nobody asked you to come here," and stepped back and made as if to land a blow.

He could never remember afterwards exactly what happened. The next thing he knew the boy was floundering waist deep in the stream. He himself must have thrown him there. At least that's what Ed Ware said. It was Ed who came pushing through the willow boughs. "For God's sake!" he said. "You Bowery bum! I turn my back, and you start a fight."

"I'm not fighting," he said.

"Oh, no," Ed said. He turned to the girl. "I'm going to take him back to Eighth Avenue. I won't let him get out again." Under his breath he added, "He's her *fiancé,* you dope!"

He had not realized until that moment that the boy had had any right to object to his being there with the girl. He told her that he was sorry. The boy was up the bank, coming at him again. He was in the act of apologizing to him, too, when Ed jerked him away up the path. His collar and tie were torn. Ed said he could not show himself at the front of the house. They went in by the servants' entrance and Ed telephoned for a cab to come out from town and take them to a train.

The next day he telephoned the girl and apologized again for his conduct. She did not sound as if she were angry with him, and he got the idea that she did not take her engagement very seriously. They had dined together that night, and the next night . . . Three weeks later he had told Mrs. Sewell that he wanted to marry her daughter and that her daughter was willing to marry him. She said that Catherine was too young to marry anybody, and pointed out the difference in their ages, nine years and eight months. He had not said anything to that, just sat there smiling at her. She had her hands clasped in front of her, a habit of hers when she was talking earnestly. Suddenly she flung them wide. "I climbed out of the upstairs window to marry Joe," she said. "My mother thought we were both fools."

He got up and stood beside her and put his arm about her shoulder. "It will be all right," he said. "I promise you it will be all right."

But he had not had any idea how he could support a wife. At that time he had never thought of becoming a professor. Bob Upchurch put the idea in his head. "If you want to get married, why don't you get a job," Bob said. "I bet old Briggs'd fix you up."

Old Briggs had fixed him up, with the instructorship that last fall had turned into an assistant professorship. He had always said that he would never teach history for a living, but he had been doing it now, for fifteen years.

VII

THE DOORBELL RANG. He did not move. It rang again, peremptorily. Evidently whoever was there knew that he was at home and was determined to get in. He opened the door. Catherine's friend, Molly Ware, the wife of his old college friend, Ed Ware, stood before him.

She said, "Hello, Jim," and walked past him into the hall.

He followed her into the living room. She had taken a seat on a sofa. She was, as usual, smartly dressed, in some sort of dark thing, with a bright, golden feather curved high on her hat. She had been in his thoughts a moment ago, a young girl. She had been so pretty, with her dark eyes, her creamy skin. It occurred to him that anybody who had known her then would not have recognized her now. The magnolia skin had gone sallow. The brown eyes were ringed with shadows that at times made them look owlish. But one could not forget that she had been a beauty. It spoke in the carriage of her small body, in the assured way in which, leaning over now to put out a cigarette, she fixed his face with a long, level glance.

"Have you heard from Catherine since she went down there?" she asked.

He stared at her. "Will you have a drink?"

"I don't want a drink," she said. "Yes, I do . . . Jim, I'd like to break your neck."

She uttered the words through clenched teeth, and he saw that as she stood there at the door she had been weeping.

Catherine and Molly had been friends since they were little girls. He said, "I'll get you a drink," and walked into the kitchen.

He opened the door of the refrigerator. Inside four miniature ice-bergs rested, apparently on air. He touched one of the ice-bergs with his finger. It did not move. There must be some implements somewhere in this kitchen, hammers, knives . . . When he and Ed Ware and Bob Upchurch roomed together on Brooklyn Heights Ed used to hide the kitchen knives every night because he, Jim, walking in his sleep, had once chased Ed with a knife . . . Directly below the refrigerator trays some crystal clear, gelatinous substance, oblong in shape, protruded from a yellow bowl. He considered inserting it in Molly's glass in lieu of ice, rejected the idea and, slamming the refrigerator door shut, found two glasses and filled them full of whiskey mixed with water from the tap.

She accepted the drink and a cigarette that he found for her in a box on the table. She lifted the glass to her lips and made a childish face. "You might at least have put *cold* water in it," she said.

"It's bad for your stomach," he told her absently. There were no more cigarettes in the box, and he had smoked the last one from the packet that he had had in his pocket. He saw another box on a table in the corner and walked towards it. *Down* . . . Molly had said. He did not at this moment know where his wife was and yet, during these three weeks, he had seemed to know. On the West Coast. The morning of the day she left they had been reading, at breakfast, a letter from Helen Tyrrell, a girl Catherine had gone to school with. Helen had just landed a small part in a Goldwyn production and wanted Catherine to come out for a visit. They had discussed whether they could afford for her to make the trip. He remembered disguising his feelings as they talked. He had wanted her to go. He had wanted to be here in this apartment, alone. That was because he had thought that he was in love with Edith. It strains the nerves to live in the house with one woman and be at the command of another, unless you are a man who delights in secrecy, in betrayal for its own sake. But he was not that kind of man. He had thought that he was in love with Edith. There had been times in the last few months when he had actually flinched at the sound of Catherine's voice, had wished that for a while, at least, he could be separated from her . . . But finding that note on the typewriter had been like a blow in the face. Since then he had not been able to get his mind off her. In his mind he had accompanied her on her journey, as he thought, to the Coast. There had been one night when he could not sleep, seeing over and over a wet, black road, a car, skidding from it, to crash in a ravine. On the fifth day he had said to himself: She is there now, and safe. If anything had happened to her I would have been notified. Her driver's license, the other papers she carried in her purse would identify her . . . But *down* . . . Molly would not have said "down" if it had been the West Coast.

The silver box was empty. Behind him Molly spoke:

"Have you forwarded any of her mail?"

"No," he said.

"That's why she hasn't answered any of my letters," she said angrily. "I wouldn't even have known where she was, if Mrs. Manigault hadn't written Sarah Cotham."

Softly he replaced the cover on the little box. Mrs. Manigault, he thought. Long Island? No. Mrs. Manigault had sold that place before she went abroad to live. But, of course. The Manigaults had an old place in Tennessee, next to Swan Quarter.

He came back and sat down opposite Molly. As he lowered his long body into the chair he realized that he was very tired. He drank the warm whiskey and water. Swan Quarter. She had been at Swan Quarter all along! It was three weeks now. Those letters in there on her desk. In the morning he would put them all in a big envelope and forward them to her . . . They were not on her desk. They were in a drawer, hidden under a pile of his dress shirts. He had thrust them there last Saturday, when he came in and found Lily standing, her dust cloth over her arm, holding a letter in her hand. Her dark face, bent over the letter, had been solemn and intent, as if she hoped that a scrutiny of the superscription might reveal the whereabouts of the person to whom it was addressed . . . Lily was in the apartment when Catherine left. She must have suspected something. Suspected what? That his wife had left him? Well, if she had, he did not have to give an account of himself to a negro maid—or to this woman sitting here before him. But he had known her a long time. A more pertinacious little devil did not exist. She would not leave until she had spoken her mind. He had better let her get on with it.

He set his glass down. He said, "Molly, I realize how fond you are of Catherine—and of me, too. But had you thought —I mean you may have to be fond of us separately. You know I am ten years older than Catherine . . ."

She made a gesture of dissent, almost of disgust. "Don't

feed me any of that stuff, Jim. I know what's the matter with you and Catherine."

"What is the matter?" he asked quietly.

"You're in love with another woman. At least you're having an affair with her."

There was a silence, then he said, "Who is she?"

She made another abrupt gesture with a silver paper cutter that she had taken from the table beside her. "I don't know—and I don't care. It's been perfectly obvious for weeks, to anybody who took the pains to look."

He thought of the woman to whom she referred, a tall, blue-eyed girl, who, like him, came from a small town in Ohio and taught history for a living. She lived in an apartment house on East Seventieth Street. For several months he had been going there to see her, secretly. They had once spent a week-end together, in Boston . . . But he could swear that Edith Ross had never met Molly Ware or any of her friends. He was annoyed to find muscles in his cheek tensing. The flesh itself felt brittle, as if defenseless to a thrust that she might make with the paper knife. It was all the mask that he—or any one else—had, and yet, as she said, it could be penetrated by a second sharp glance.

"It's possible that my—peculiar conduct may be due to something else," he said. "Have you ever considered that my work makes extraordinary demands?"

She shook her head. "Don't talk like a professor," she said, and added contemptuously, "Your *work!* Jim, I may be dumb—in fact I know I am—but I'm not dumb enough to think it's your work. Bob says that's the trouble, really."

"Bob?" he asked, frowning.

"Yes, *Bob*," she repeated, pronouncing with a defiant ring the name of the friend with whom they had both once been so intimate and whom they now rarely saw. "We were all at Pete's the other night."

He leaned back in his chair and, crossing one leg over the other, carefully adjusted the crease in his trousers before he said in a faintly sarcastic tone, "And just what does Bob think is my trouble?"

"Oh, he says it's all too easy for you. You can run rings around the ordinary professor, and yet there's just enough work to take the edge off your mind . . . He says you'll probably never write another book."

He was silent, turning his head with a quick motion so that he gazed past her out on to the street. Seeing his profile against the white blind she remembered how, when as a young girl she had first met him, she had thought, seeing the big nose and the lock of hair jutting over his forehead, like the crest of some bird, that he was one of those men whom you could describe as "hawk-like." He had seemed to her fitted for bold, decisive actions in which he would not stop to reckon the consequences. In her provincial ignorance she had been surprised when they told her that he wrote books.

He was turning his face back towards her. "A moral collapse, eh? Well, Bob ought to be an authority on moral problems."

She felt tears springing to her eyes. She lowered her gaze to the floor and blinked her lashes. "Bob hasn't had any moral collapse," she said dully. "He's just gone on, the best he could. I know you think it's filthy in him to get three hundred a minute or whatever it is over the radio, but I don't see that it would help things if he refused. It's the set-up. And he has to make a living."

"I'm not suggesting that he stage a one-man revolution," he said in a tone that he had never used to her before, bland but acrid. "Perhaps the fault lies with his sponsors. Duke and Crosby—or is it Bates and Sanborn?"

"It's ci-cigars," she said, "Manuelo y—something."

A tear rolled down her cheek. She shook her head angrily and stood up. "I've got to go," she said.

He was on his feet. She knew that in order to show that there had been no breach between them he would insist on accompanying her downstairs and finding her a taxi. She put her hand out, palm up, in a gesture of refusal. "Don't bother," she told him. At the door she turned and spoke again:

"If you don't care anything about yourself you might think about your friends. Think of Ed, sitting there, night after night, writing those heads and then going out and getting drunk. I used to fuss but now I know he couldn't stand that damn desk without it. But the doctor says his kidneys . . ." She was crying now, into a handkerchief that, with shaking hands, she had rummaged out of her bag. He knew that he must not go to her and stood, silent, his eyes on the floor, while she wiped her eyes. She finished and, stuffing the handkerchief back into the bag, lifted a streaked face in which the brown eyes, ringed by fatigue and emotion, looked more owlish than ever. "You *ought* to care," she said, "he just adores you," and slammed the door.

He heard her footsteps in the little hall and then the outer door closed. He went to the table and, taking up his glass, drank slowly, then set the glass back on the table and, going to the window, stood looking down at the street. *Adore,* he thought. A queer word to use of such an ungainly, lonely man as he was. There had never been anybody who adored him. That girl, Edith Ross, was in love with him now, he supposed. But she could just as easily fall in love with somebody else. With Wilkins, for instance. The women all admired Wilkins . . . Swan Quarter . . . So that was where she had been all along. It was strange that he had not known it. But why had she gone there? She had a kind of dread of the place and sometimes, walking along Fifth Avenue, would hurry into a shop to buy fine woolen garments for the old lady, a fleece-lined jacket for the old maid aunt. You don't know how cold it is, huddling over grate fires . . . Aunt Willy is out in all kinds of weather, looking after the stock . . .

They had made a visit there, once, soon after they were married, at his suggestion. Catherine had not much wanted to go. They had driven down and as they passed through the Lewis' market town of Carthage, Catherine had insisted on stopping and loading the car with soap, loaves of bread, cheese, heads of lettuce, even a great roast of beef. "We can drive in any day and buy anything we want," he had told her. "I know, but something might happen. The car might not go. Things always get out of fix at Swan Quarter."

The two women, who did not own an automobile and could not have driven it if they had, lived in great isolation. Their only communication with the outside world, except for chance visits from relatives and neighbors, was through an old negro man who, twice a week, drove in to Carthage for supplies. And yet the place was not as run-down as he had expected. The house needed painting and there was no plumbing, but he had lived happily in much more primitive quarters in Europe. It was the isolation that gave the place its melancholy aspect, not far from the highway but deep in a wood and circled by water; the creek made a great bend that took in all the land except a ridge next to the Manigault place. When Catherine was a child she had always gone back there in the summers. Had the isolation that she had known there set its mark on her? Her mother, though she had lived in New York for fifteen years, always seemed to have that old farm in the back of her mind. On a February night she would walk to the window and look down on the lights and the river, saying, "The jonquils are blooming now at Swan Quarter," or, "Willy's nursing twins now." Ewes, when they bore twins, often refused to suckle one which then had to be raised by hand. Catherine still talked about an orphan lamb that she had had for a pet.

He imagined his wife there in her room. Over the mantel hung a picture of a great-aunt dead in infancy. The child, who had died long ago on a summer day of a surfeit of ripe peaches, had looked down all through her childhood on the child Catherine.

The house he had grown up in had been sold soon after his mother's death. He had been abroad when his mother died and had desperately needed the small amount of money that would come to him from the estate; so the house, a white frame house on a quiet street, had been sold. He had not thought that he would ever go back to Mount Hope; so he had sold the furniture too, reserving only the books. The friend of his mother's who had packed the books and shipped them to him had included one or two things that she had evidently thought he would want as keepsakes: a cushion that had lain in the rocking chair in which his mother always sat beside the sunniest window to sew, some old notebooks of the doctor's, some family photographs of whose existence he had hardly been aware, the family silver. He had not remembered until he looked over the things how many objects go to make up a household. His wife sometimes used the silver, his father's and mother's photographs sat on his dresser, but the rest of the things, packed in a long box, were stored in a warehouse here in the city. "But I might carry them about with me," he thought, "like the boxes of his native Carpathian earth that Count Dracula took with him to London."

He had not sold his father's books. They were ranged on a shelf on the east wall. He went over and took one out: a dark, worn copy of *The Decline and Fall of the Roman Empire*. A great man for Gibbon, his father. He would come in from a late call and, taking off his shoes, sit and warm his feet over the hot air register and read far into the night. And the next day he would talk to you about what he had read; he had the faculty of making it seem contemporaneous. He could remember now the old man's indignation over the figure Heliogabalus cut when he entered Rome: in silk robes, his cheeks painted, his eyes mascaraed like a girl's. And the senator Julian! For some reason he had thought very highly of Julian and grieved when he accepted the imperial crown. "Couldn't he see that he was signing his death warrant?"

A tall, lean man, the doctor, with a sharp chuckle and an air of dispassionate benevolence that never left him. "Well, Joe, if I were you I believe I'd . . ." He would hook his thumbs through his vest and with his head, sparsely covered with graying hair, a little on one side, calmly and kindly consider what he would do if he were Joe Sienckewicz and had a wife and four children and an incurable disease. He had lived in Mount Hope ever since he first came there to practice, a young man of twenty-four. But he never seemed to think of it as home. "The people here . . ." he would say or, "In this part of the country . . ." Once, when Jim, a boy of ten, had accompanied him on a country call he had driven a mile off the road, to the top of a high hill. They had sat there awhile, looking off over the country. Before they left he leaned forward and pointed with his whip to a faraway, green eminence. "That looks a little like the Green Mountains," he said.

He was probably always wanting to get back to the Green Mountains, his son thought. I don't feel that way about Mount Hope. But perhaps that feeling of his kept us from ever really belonging in Ohio. Or was it something in the soil itself? He had heard his father commenting, with his dry chuckle, on the way the Ohio farmer who had accumulated enough money sold his farm and came to town to live out the rest of his days. A Vermont farmer, he said, would stay by his land till he died and then have to be carried off feet first.

Was there something in the land itself that repelled human attachments? Perhaps it was too fertile. Roots put down easily are not as enduring as those which make their way through the interstices of rock. The richest, most persistent civilizations have flourished on the barren slopes of the Mediterranean, where man has to conserve every inch of soil. He had heard his Provençal landlady exult over the death of a twenty-seven-year-old horse. "It had arrived for him to die. Now he will help make next year's garden." Middle Westerners, springing out of the rich, black loam of the prairie, are always on their way somewhere else . . . I never felt at home but once in my life, he thought. And that was in that room over on Eighth Avenue.

The telephone rang, once and then again, a long, insistent peal. That was Edith, he thought, and did not go to the telephone, but instead, crossed the room and sat down. Flinging himself back in the deep chair he put his hand up to his forehead, pressing his eyes shut with outspread fingers, as if the same fingers that shut out sight could shut out sound. The room on Eighth Avenue had been distempered pale blue, but long ago. Soot and grease and the occasional overflow of bursting pipes had laid strange stripings of green over the blue. Bob Upchurch had once taken an eraser and rubbed a small section of the wall clean, leaving the green streaks thrusting like giant, Rousseauistic ferns to the ceiling: T. S. Eliot's rat, he said, creeping softly through the vegetation.

There was an iron bed, so frail that it creaked when you turned over, a table and two chairs, one for him to sit on while he wrote and one, as Bob said, "for solitude." There had also been a shelf, high on the wall, where he set his frying pan and dishes and what food he had on hand. The little stove had had only one eye and would not hold a frying pan and a kettle of water at the same time. The coal that fed it was kept in a big box outside in the hall, along with the kindling and some old newspapers. Making the fire on winter mornings was the worst part. After he got it going well the room would heat up. But only for a time; the fire had to be replenished every hour. But he would forget that when he got to writing, forget even that there was a stove until, after midnight, when the room would grow colder and he would get up, and with his eyes still on the words he had just written, take his overcoat down from its hook and in that brief pause, the only one that he might know before daybreak, become aware of the silence that had settled over the city, a silence that in those small hours used to grow and deepen until, miraculously, it found a voice and spoke to him, his constant, dear cell-mate, the best companion he had ever had. He replaced the book upon the shelf and turned, listening, as he gazed about the room. But the silence, after all those years, gave back no sound.

VIII

When he had come out of the apartment building he had walked east and then, at Third Avenue, had turned downtown. And for several hours he had been walking beside the Elevated. It was dark overhead; lights were burning in all the shops. Back there it had been junk shops; now it was delicatessens and fish. There was something familiar in the way that eel coiled itself about a heap of shell-fish. He stepped up to a window and looked in. A miniature castle reared itself in a huge abalone shell in the corner, a small turtle peering gloomily out from under its draw-bridge, and back in the sawdust three dark, white-aproned men sat playing cards around a backless chair. He had come here once, with Catherine, to buy mussels. An extremely cold day in winter. One of the men had looked up briefly. "No mussel—no clam—no dig this weather," he said and jerked his thumb to the east. Catherine, not able to connect the ninety-mile gale that for three days had beaten the Atlantic coast with this shop, had been at first incredulous, then annoyed.

She had six people coming for dinner and had promised them *moules marinières*. The party had been for Ellen Page. He had never liked Ellen but later he had been glad to have somebody in the apartment besides himself and Catherine. The dinner party had been on Saturday night, the night before he had seduced Edith Ross. Friday—or was it Thursday that he had gone to her apartment to work on those papers?

Where you going, Mister?

Chapman stepped back quickly, then moved aside. The grimy hand fell from his arm. He walked on down the street.

Won't the gentleman give me something for a cup of coffee?

"No," Chapman said and walked faster.

An odd COFFEE POT sign, that, with the red and green dwarfs leaping over the letters. The one that leaped from T to E had less work to do than the one that leaped from E to C. If you turned your head you could see what they did when they came to the end of the letters, and if you wanted something to eat you could cut over to Madison or Lexington. I don't want anything to eat, but I could do with a drink. *Seduce.* That was the word, all right. She never made any play for me. That time in my office when we were talking and she suddenly stopped and walked to the window and I thought, "My God, she's thinking the same thing I am!" but she said afterwards that she wasn't, that such a thought never entered her head. Not then, she said, but another time, when we were standing on the street and talking and Wilkins came along and said something about her hair, the way the wind was blowing it, and she saw me looking at it, she did think about it then, she said. But it wouldn't have happened if I hadn't gone to her apartment that night.

Catherine tried to keep me at home: the Wares were coming in after dinner. She was kneeling by the fire, doing Ellen's hair a new way before they came. It doesn't make any difference how they do Ellen's hair. Her face always looks like a hickory nut. Catherine likes women like that. She had hairpins in her mouth. She didn't even look up. "Oh, he works almost every night. And he has the most attractive girl to help him. Her name is Ethel Ross."

I said, "Her name is Edith. She's pretty damn good." It was raining when I got out on the street. She had a fire, too. The bricks in her hearth are yellow, not red like ours. She had on one of those long, soft dresses that they wear in the house. She never puts any hairpins in her hair, it's so short. It was after we had finished the papers and she went into the kitchen for the bread and cheese. I followed her. I laid my hand on her arm. She turned around quick. She said, "Jim, what is it?" I knew then that I still had time to get out of it. But I didn't want to.

He had stopped and was staring into a window where a dozen bottles were pyramided on a *papier maché* revolving stand. "I didn't want to," he said aloud. *Coon Hollow* spun slowly out of sight and was succeeded by *Old Taylor*. He became aware that he was not alone. The bum was standing a foot away.

"God, you stink!" Chapman said. "Is it dirt or that stuff they disinfect you with?"

The man looked at him, then his eyes went back to the window. "I can't help it," he muttered. "Charley won't let me sleep in the house."

"I don't blame Charley," Chapman said.

The man slowly took his eyes from the revolving bottles. He shuffled closer to Chapman. The acrid odor, stirred by the movement, rose and hung in the air between them. The bum's face showed red through grayish bristles. His eyelids, gaping under pale blue eyes, showed a deeper, almost arterial red. "You go to the Army," he said, "they put your stuff in the machine efery night. Don't make no difference how often you come. They put it in. Efery night."

"Salvation Army?" Chapman said absently. "I suppose they've got to get the lice out of 'em some way . . . Why won't Charley let you sleep in his house?"

The man shrugged his shoulders.

"Thinks you stink, eh? How come he's so particular?"

The bum shook his head. "I don't know. He a colored fellow." He shuffled closer. "Won't the gentleman gif me something for a cup of coffee?"

Chapman drew back. "I'm not a gentleman," he said, "I'm a scholar . . . Yes, I'll buy you a drink."

They entered the bar. "Two double Scotches, straight," Chapman told the bartender. The bartender turned to the shelf behind him, then paused, his hand on the bottle he had been about to take down. Chapman laid a bill on the counter. "This is a friend of mine," he said. "You want to give him a drink, or do I take him to some other bar?"

The bartender leaned forward. "Mind taking him over there?" he said in a hoarse whisper, and pointed to a booth in the far corner.

"Not at all," Chapman said. "He prefers to sit down. Come on, pal."

They walked over to the booth. A boy brought the whiskies. "You'd better bring us something to eat, too," Chapman said. "What'll it be, pal?" He turned to the boy. "What have you got?"

The boy, staring over his head, indicated with a gesture

the menu stuck behind a metal napkin holder. Chapman unfolded the greasy menu and read: *"Spaghetti, with meat balls . . . Chili . . . Mexican Red Hot Tamales . . . Irish Stew* . . . What'll it be, pal?" he repeated impatiently.

The bum's mouth, half open, showed pale gums, rotten, broken teeth. He murmured "Chili," then shook his head. "Spaghetti," he said. "Spaghetti and meat balls." Closing his lips firmly he fixed Chapman's face with his watery blue eyes.

"One spaghetti and a ham sandwich," Chapman told the boy. "And you'd better bring two more whiskies."

The bum, his eyes still on Chapman's face, stretched out a blackened hand and took potato chips from a bowl that stood between them. Chapman pushed it towards him. "That's right," he said. He lit a cigarette and smoked, staring past the man, past heads showing over the tops of other booths to where the street lights showed dimly through the cloudy window.

The boy brought the order. The bum, after another glance at Chapman, began eating. Chapman smoked, letting his sandwich lie untouched.

The most attractive girl. Her name is Ethel Ross . . . She said that. It hadn't happened then. But she said that. She *made* it happen. No. She is just a flighty girl. But you are a son of a bitch. A fool, too. You always were a fool. Bob Upchurch said so. Jim, you've got something like genius, but you sure are a fool. But that was when we were kids, living in that room over on Columbia Heights . . . It's too late to do anything now. Still, you know where she is. You could write her a letter. Tell her you are not going back to Edith . . . Not going back. I'll write *her* a letter. No, I'll have to see her. It's my fault, Edith. All my fault. But it will be better to break this off now. Better for both of us. My wife . . .

Don't bring me into it, Mister Chapman. Don't let me stand between you and Ethel . . .

Her name is Edith. She wants to get married . . .

I don't blame her, Mister Chapman. You are such an attractive man, and you can be so persuasive when you put your mind on it. Tell her I won't put a stone in her way. Not a stone . . .

It's a straw. You talk too much and mix your metaphors.

Now, isn't that too bad! And after fifteen years of associating with you! But I am only an ignorant country girl . . .

You are selfish and cold-hearted. You never even knew all I gave up for you . . .

It was a mistake. If I hadn't been twenty years old at the time and green as grass I'd have advised against it. My heart bleeds when I think of the sacrifice. You ought never to have married, Mister Chapman. You ought to have been like that king and had mistresses. You can have a lot now. Ethel won't care . . .

Her name is Edith. Edith Edith Edith Ross. She wants to get married. I'm not going to marry her, he thought, surprised. I'm not going to marry her.

The bum speared up the last strip of spaghetti and rolled it about on his fork. Chapman pushed the sandwich towards him.

"This Charley," he said, "is he a bum too?"

The man shook his head. "They pay him," he said. "They pay him, but it's Hans takes up the ashes. Efery morning. At seven o'clock. Two Hundred and One West Eleventh. Hans," he said vaguely, "Hans Hofschwengel."

"Let's have another drink," Chapman said. "You've got a job then?" he said, when the boy had brought the drinks. "Charley gives you something for taking out the ashes, doesn't he?"

"He gifs me bed money efery night," the bum said, "and he gifs me a quarter. Sometimes he gifs me a quarter."

"You born in Germany?" Chapman asked.

Hofschwengel nodded. "Düsseldorf," he said. His pale eyes lighted. "But I go to the university at Leipsic. You ever gone to Leipsic?"

"Many times," Chapman said. "Well, you must have learned a lot at the university."

The man's glass was at his lips. Over its rim his eyes, grown suddenly intent, sought Chapman's. *"Non vitare, plagas in amoris ne iaciamur . . ."* he muttered. His gaze wandered, fixed a point to the right of Chapman's shoulder, then swung back.

> *"Non vitare, plagas in amoris ne iaciamur*
> *non ita difficile est quam captum retibus ipsis*
> *exire et validos Veneris perrumpere nodos . . ."*

the bum chanted.

Chapman laughed, then gazed into the man's face. "You're right—it's easier to get into the toils of Venus than to get out, once you're caught." He laughed again. "I always preferred a later, even more pagan version of that idea."

He rose and, plunging his right hand into his pocket, brought up a handful of change. He stood silent while the man slowly counted the coins, then, suddenly, he set his hand upon the damp, greenish shoulder and leaning so close that he could smell the fellow's foul breath, recited:

"Cras amet qui nunquam amavit quique amavit cras amet."

Hofschwengel blinked. "I don't understand so good," he murmured.

Chapman laughed. "I never did either," he said and made his way through the smoky air to the bar and drank another whiskey. He put the glass down and went out into the street.

IX

AT THE CORNER OF Third Avenue Chapman gave a sleeping taxi driver the number of his apartment house on East Sixty-fifth Street. The cab moved away under the long shadow of the El. Chapman, watching the black columns flash past, remembered that a woman had once told him that in childbirth she had had a dream of an elevated railway, whose rails, instead of running parallel, diverged. She and her husband, mounting, for a lark, into the little saddle-like cars with which each rail was fitted, had realized only when they came abreast of each other and saw the rails spinning before them into space, that they were bound for infinity.

"Get over on Madison Avenue," he told the driver.

The man crossed over and was starting uptown again. At Forty-fifth Street Chapman called out to him to halt.

"I'll get out here," he said, and alighted in front of one of the little green hedges, which, springing from tubs or boxes set on New York pavements, attempt to re-create for the nostalgic refugée or former expatriate the atmosphere of the Parisian café. A man was just disappearing under the red and green checked awning. The figure, glimpsed from the cab window, had seemed, from the gait and swing of the shoulders, to be that of Bob Upchurch. He had not seen Bob for a long time, but earlier in the evening—or had it been a day or two ago?—he had been talking about him, with somebody whose identity for the moment eluded him.

It seemed to him that he had to see Upchurch, and, settling his hat more firmly on his head, he strode into the café.

It was full of people but he did not see Bob or anybody else he knew. He studied the faces around him, then pushed his way to the bar and ordered a double Scotch. The bartender, as he set it before him, leaned forward, pointing. "There's a gentleman over there wants you, sir."

Chapman turned and saw Ed Ware, half rising and beckoning to him. Ed was wedged behind a table in a corner of the room. The man sitting opposite him stood up as Chapman approached. It was Upchurch. Ed was sliding the table out. Chapman slipped behind it and sat down on the red, leather-upholstered bench that ran the length of the wall.

"It's a good thing you brought your drink with you," Ed said. "Takes pull to get anything here tonight," and, standing up again, he began to bellow in an anguished tone: "Mike . . . OuAH, *Mike*!"

The black-haired barman, hanging over another table, inclined his heavy body towards him in the immemorial gesture with which waiters, while remaining motionless, try to convince you that they are flying to do your bidding, at the same time asking indulgence by a levelled forefinger. "Club, with chicken!" Ed yelled, and, sitting down, picked up his drink. "I've got to get out of here in half an hour," he said. "Jim, where you been keeping yourself?"

Chapman, as he was making his way towards them, glass in hand, had lurched, and recovering his balance had realized that his head was spinning. He pushed his glass a little way from him and spoke slowly, formally.

"I was in Maine most of the summer."

"Must be fine to be rich," Ed said. "I'm telling you, it was a beaut here this summer."

"Didn't you . . ." Chapman began and, pausing, recalled with difficulty the fact that Ed's wife was named Molly. "Didn't you and—Molly get away all summer?"

"Far as Fire Island," Ed said. "We're short on the desk now."

Bob Upchurch put out his hand. "How are you, Jim?" he asked.

"I'm fine," Chapman said. The disagreeable spinning sensation was gone, replaced by an airiness that carried with it a conviction of complete lucidity. He had the impression that he was standing at a little distance from the table and yet his senses registered with unusual precision all that went on. Listening to his own voice it seemed to him that his enunciation had the roll of drunkenness, so he repeated the words until he succeeded in pleasing this new, critical self. "I'm fine. Feeling fine."

"You look it," Ed said with a laugh.

The waiter brought Ed's sandwich and two drinks. Upchurch pushed his away. "I ordered a Tom Collins," he told the man.

Ed was taking his sandwich to pieces, lifting out little pieces of sliced olive with a fork. "I don't like to eat olives this late at night," he said.

"That your supper?" Chapman asked.

Ed lifted his worn, saturnine, incurably good-humored face. "Sometimes I have a steak," he said, "but usually this late, I just take a sandwich." He bent his head for a quick bite, at the same time reaching for his glass. Chapman observed that his forehead, even his sunken cheeks were lightly dewed with sweat. Ed went to work at eight o'clock, after an early dinner. At midnight he knocked off to eat this supper and in a little while he would go back to the desk to rewrite late stories that might be telephoned in. He would have another drink, maybe two before he left. He had once told Chapman that there was no use in his abstaining because he was working. "I can write a head just as well drunk or sober," he had said. Ed came originally from Jessup, Georgia. He had worked on the *Atlanta Constitution* for ten years and then, still in his twenties, had come to New York. Chapman recalled the day that he had been promoted to a desk on the *Globe*. "He'd have been better off if he'd stayed a leg-man," he thought and wondered how long it would take the desk to kill Ed. The agent of destruction would be the light film of fatigue which glistened now on his features, like down on the skin of a peach. It would strike deeper all the time, until, filtered through bone and flesh and blood, it solidified, the hard, unyielding core of his being.

The waiter brought Upchurch's Tom Collins. Ed looked at it irritably. "What do you want to drink gin for, Bob? It'll kill you."

"I got in the habit when I was lecturing this spring," Bob said. He raised the glass to his lips a moment and set it down. "It's colorless," he explained, "and there's always some Junior Leaguer who'll slip it into the pitcher for you."

He turned his head to watch some newcomers, revealing a remarkably handsome profile. Chapman stared at him.

When Bob was eighteen he had looked like one of Donatello's boys. A full-orbed, hazel eye, a high forehead, crowned by sculptural curls, an aquiline nose subtly curbed by the well-cut nostril, a mobile, generous mouth. The hazel eyes were the same but the layer of flesh that comes with middle age had somewhat obscured the other features; the flanges of the nostrils were heavier, the lips thinner, less sensual.

He had observed these changes before but tonight it was as if he were seeing them for the first time. It occurred to him that some individuals have their true identity only in youth. All growth, all subsequent change seems to have been made by some disease or disorder of the blood, which subtly alters the features from their true mould. He leaned forward and fixed his blood-shot eyes on Upchurch's face.

"What were you lecturing on?" he asked.

"'Some Contemporary Trends in Fiction,'" Upchurch said. He smiled and, raising his glass, brushed it with his lips and held it out as if for a toast. "Or What Songs the Cash Register Sings."

Chapman, his head a little on one side, seemed to ponder the title. He was recalling with surprise that Bob, who every week wrote a critical article for the magazine, *Procession,* at what seemed to him a fabulous salary, was considered an authority on the contemporary novel. In his youth he had been a poet.

In the big room on Columbia Heights that they had shared that last year, he had read poetry to them: his own, his friend's, Marvell's, John Donne's, John Webster's. He read poetry perhaps as well as it can be read. Syllables, dropped by his clear voice into the smoky air, seemed to be illuminated by their own inner meaning. Sometimes the rhythms that had formed in some man's mind three hundred years ago beat through the open windows out onto the city streets. One night the poet from 110 pushed the door open and taking a folded paper from his mackinaw silently handed it to Bob, and Bob read:

Mark how her turning shoulders wind the hours,
And hasten while her penniless rich palms
Pass superscription of bent foam and wave,—
Hasten while they are true,—sleep, death, desire,
Close round one instant in one floating flower . . .

A fellow named Hart Crane, who worked for an advertising agency by day. His poetry had just become famous when he leaped to death from the stern of a Caribbean steamer.

In Italy, from a dirty, three weeks old *Times,* picked up at a café table, he had learned of the suicide and had sent Bob a cable. Bob's letter was in his room when he got back from the American Express Company. "A good thing, Hart's death," he had written. "He was through, and knew it." But he had taken the title for his own volume of poems from Crane's work: "This Fabulous Shadow"—not the shadow from Plato's Cave, but Crane's line, "This fabulous shadow only the sea keeps." He had used to know most of those thirty poems by heart. Bob's book had come out in 1930. There had been no other book since then. That damned busy-body, Henry Etheridge, had got him the job on *Procession* next fall.

They were still talking about the lecture trip. ". . . I then dissected another current favorite. She was a composite personality. At one time we had Charlotte Brontë, Ethel M. Dell, William Makepeace Thackeray and Thomas Nelson Page scattered over the rug. . . ."

Chapman listened, the brooding expression still on his face. It was the voice, he thought. Clear, carrying, well-modulated, strongly infused with sexuality. When he talked he bent his eyes full upon you and he seemed, if you were talking, to be always on the verge of giving you his swift, flashing smile. Crane, a stubble-haired, pop-eyed fellow, who seemed to live only for poetry and had ended his life when it failed him, had had a jerky, nervous voice and was an insufferable egotist.

Bob was too good-looking, too smart, knew his way about in the wilds of New York too well, a rat creeping through the vegetation, no, a stoat that, padding in winter to the water-rat's burrow, changes its coat from red to white, so that it may not be seen above the bank. But the trapper, taking him in the snare, wants only the white, soft fur and, rending the fierce cords, throws the brittle brain-pan, the sanguinary eye, the subtle guts back on to the snow.

". . . My agent goes ahead. He demands that the stage be draped in black and that wreaths of calla lilies, interspersed with myrtle, be disposed about it at intervals. I am, he explains, in perennial mourning for the printer, Samuel Richardson. The secretary of one club, a very attractive young matron, wanted to know more about the deceased printer whom I mourned, so we went up to my room and were having a nice talk when the president of the club, Mrs. Horace J. Dellenbraat, appeared, saying that the mayor wanted to see me. I was afraid that I had broken one of the city ordinances but it turned out that he only wanted to give me the keys. Very nice fellow. Went to Princeton. We all drove out to his house after dinner and played 'Monkey Pile.' . . ."

In ancient times, Chapman thought, the wearing of ermine was restricted to royalty. So, a poet might be given to a prince, alternately to pet and to torture, as Torquato Tasso was given by his father to the d'Estes. Now the mob is privileged of them.

"Bob, you don't read all those books you talk about?" Ed asked curiously.

Upchurch nodded. "It is turning *me* into a composite personality. I am both Bouvard and Pecuchet. And they are splitting. Bouvard is interested in the historical novel. He read fifteen last month. Pecuchet is for *sur-realisme*. Takes to Henry Miller."

"If I was going to write a novel," Ed said, "I'd take a high class whore-house and I'd tell the club ladies how the place was run, what kind of soap they used in the bathrooms, what the maids said when they called the girls down . . . I'd give 'em a few glimpses, of course, enough to make 'em feel they'd really been there."

Upchurch shook his head. "That was Defoe's formula. Won't work now. You'd have to put in some chi-chi, not the old gag about how the madam was the best woman in town, but some long, Proustian sentences. Turn her into a seeress. Look up Erskine Caldwell if you're interested. But I don't believe you could make it, Ed. You'd have to be pretty vatic." He yawned. "If I wrote a novel," he said, "it would be the life story of the lobster that Gerard de Nerval used to lead around on a string. It neither barked nor bit and knew the deepest secrets of the sea . . ." He looked at Chapman. "Jim, are you writing anything now?" he asked.

Chapman, his head still turned slightly aside, and supported by his hand, had been trying to recall a poem written by the drowned poet. He had got as far as

> Where icy and bright dungeons lift
> Of swimmers their lost morning eyes . . .

but an older refrain beat up and submerged the lines.

He turned a heavy, melancholy stare on Upchurch. *"Perdidi Musam tacendo,"* he said, *"nec me Apollo respicit."*

"He has lost his Muse by being silent," Ed offered in an appropriately formal tone. "Apollo no longer regards him." He took a last bite of sandwich. "Never was a boy went to Webb School flunked in his college Latin," he said. "You know why? Old Sawney beeched it into us. Let a whole grove of beeches grow up convenient to the house." He laughed. "Was one old boy named Taylor. Sawney asked him what a taxidermist was. 'It's a fellow that collects taxes,' Taylor told him. Sawney pretty near cut down the whole grove that day . . . Jim, didn't Kit's uncle go to Webb? Sort of wild fellow, named Jack Lewis?"

Chapman did not answer. While Ed was talking he had raised his head to meet Upchurch's eyes. Through those bright hazel pools curiosity had flickered and then was gone, as swiftly as a trout's tail lashing through water.

"He's wondering about me," he thought.

Lowering his eyes he drew his glass towards him and drank.

Ed picked up his empty glass, looked into the bottom reflectively as though he contemplated ordering another drink, and set it down. "Old Sawney was death on liquor," he said. "Boys used to bury their jugs in the ground and then suck it up through a straw . . . Got so nobody'd play football or baseball. Everybody wanted to play mumble-the-peg. Old Sawney'd sit over on his porch and watch 'em mumbling the peg and it got him worried. Said it was a good, Christian game but no use being intemperate about it." He stood up abruptly, holding out his hand. "I'm late," he said. "Sure has been fine, seeing you fellows."

Upchurch stood up too. As he shook Ed's hand he laid his other hand on his arm affectionately. "How about having dinner some Thursday? That's your night off, isn't it?"

"Tuesday," Ed said. "I'll phone you. Hell, I've lost your number!"

They took little notebooks from their breast pockets and exchanged telephone numbers. When Ed had gone Upchurch turned to Chapman. "How about you coming, too, Jim?" he asked. "Make it next Tuesday?"

"I'll come," Chapman said, "Tuesday or Thursday."

Upchurch laughed. "I've got to run along," he said. "Well, good night, Jim."

"Good night," Chapman said, then suddenly rose and accompanied him. Out on the street Upchurch paused and looked at him uncertainly. "Going uptown?" he asked.

"I'm going downtown," Chapman said. "Far downtown."

But when Upchurch, with another laugh and a friendly touch on the shoulder, had left him, he did not move but stood, leaning upon his stick, and staring up at the great building whose lighted windows jewelled the dark. The buildings which flanked it on each side were not as tall but every one showed some lighted window. All over the city, people in their cubicles of stone or concrete or steel, lay as tight against one another as bees in their cells of wax, and even beyond the confines of the island the great, crowded ramparts flung themselves on and on until if one travelled far enough one might come to a building whose four walls housed one man and his wife and children.

Once, years ago, he had stood in Nîmes, gazing up at the wall of the arena, trying to imagine how it had looked when every cranny was inhabited. The Visigoths, demoralized, *Romanized,* fleeing before their one-time forest neighbors, the Lombards, had fortified those broken walls and had lived inside them, like modern apartment house dwellers. But they had lived their apian life for a generation only; when they went back to the land men who had been boys at the time of the exodus went with them, to instruct them in the cultivation of the vine, the care of the flocks.

Bees, in their solitary cells, do not control their own destinies. The young grub does not know whether the food discharged to it from the worker's stomach is the "royal jelly" that will turn it into a queen or the ordinary secretion which fates it to become a drone, which, in the winter, when it no longer has a function to perform, will be cast out of the hive and stung to death.

And the queen? O City, preparing for what strange, nuptial flight! Having stung her sisters to death, she rises on rapid wing, but when the dead bridegroom has dropped from between her feathery legs she will hurtle down, past the heaped bodies of dead and dying drones. Will not the odor of decay penetrate the royal chamber, interrupting even the processes of fecundation, so that, seeking a cleaner air, she may lead her hive forth in a last flight, in which, travelling high above the orchards and the gardens, they will not stop to cull honey from the apple blossom or the rose, but will continue on, an insensate mass, until, dying, falling in a great cloud, they darken with their wings the whole west?

X

IT WAS ALREADY LIGHT when Edith Ross awoke. Outside, in the hall, she could hear voices, one muted and thin, the other deep, and then, with an almost imperceptible click the sounds were cut off, as if they had been on a phonograph record. One of the old ladies who lived on the fourth floor was exchanging her morning greeting with Oscar before she started down in the elevator. She closed her eyes. Oscar would not have been on duty last night. His nephew, Paul, came on at twelve and stayed until seven. A pimply-faced boy who was still going to high school.

She moved over a little, so that the warmth from her companion's body struck through her thin nightgown into her flesh. Her eyes still shut, she put her arm about his waist and drew her body up so close to his that her face was pressed against his back. But the fabric of his jacket grew moist where her lips were pressed and the reverberations of her own breathing sounded loud in her ears. She feared that she would wake him and she drew a little away and, lying on her side, stared out into the room.

Near the window a man's dark coat hung on the back of a chair. Beneath the chair a shoe protruded from under a heap of linen and halfway across the room, as if the man had registered his progress by discarding a garment at each step, another shoe lay. Beside the bed, a pair of gray trousers was flung down. A pocket, gaping, showed a roll of bills.

One bill, a five, had spilled out, and lay, flat, gray-green on the floor.

She had been asleep last night when the telephone rang, so sound asleep that at first she could not make out what Paul was saying, and then something in the urgent, angry tone in which he repeated the word "gentleman" had disturbed her and she had hung up, telling herself that if it was really she who was wanted he would call again. A few minutes later there had been the click of the elevator and the discreet knock on the door and Jim, obviously drunk, stood there staring at her, with the sharp, pale face showing just behind him.

She had thought quickly, stepping past him to smile into the sharp eyes, speaking in a low, companionable voice. "It's nearly morning, isn't it? Perhaps we'd better let him come in." But it might have been better if she had refused to let him come in, had made Paul take him back downstairs, even at the risk of further disturbance in the hall. Vesta had been right in those cautions she had given her last week. "It doesn't pay to be talked about . . . A teacher *has* to be careful." Well, suppose Paul did tell his uncle? What could old Oscar do? She was a desirable tenant, paid her rent promptly and up till now her behavior had certainly been discreet. But from now on both Paul and Oscar would be watching her, every time Jim came up. She might have to move . . . She *would* move, at the first sign of impertinence from either of them. Thank God, this was New York and not Waterloo.

Jim had been quiet enough after she got the door closed, muttering Latin verses as he bent to take off his shoes, growing irascible only when she mentioned Paul. "I gave him five dollars. Hell, what's *he* grousing about?" It had been said in an unnaturally loud, distinct tone. And then, after they were in bed, his strange endearments, his harsh caresses! . . . But that was because he was tight. She had never really seen him tight before.

She had been staring through half-closed lashes at the gray-green bill. Five dollars was a lot to give Paul. But a man had to be generous in affairs of this kind . . .

She got up abruptly and, slipping on a dressing gown, tiptoed into the bathroom. A mirror was set into the door. When she had bathed she stood in front of it a moment, studying her tall, white-robed form, then took a comb and ran it through her hair. The dark, springy mass was hardly disturbed by its passage. "I wonder how I'd look with long hair," she thought and with her palms pressed her curls down on each side of her temples. But the face, framed between the dark lines of hair, seemed to her at once older and less distinguished and she swept the comb through her curls again, leaving them waving about her forehead and temples.

She went into the kitchenette. At this time of day, flooded with sunshine, it was the pleasantest of her three rooms. She put water on to boil, covered the white, enamelled table with a pale pink cloth, set out cups and saucers and squeezed orange juice into two squat, blue glasses that she had brought from Mexico last summer. The water was boiling. She measured coffee into the drip-pot, then went to the window and examined the plants she had set out in a long, green box a few weeks ago. The ivy had taken root and was growing sturdily but one of the geraniums showed yellowing leaves. She must get Vesta to look at them the next time she came to town. She had "the green thumb." The most delicate house plants throve and even blossomed for her while Edith was always having to replenish her stock from the florist. But that, perhaps, was because Vesta gave them so much time and thought, ascertaining, when she bought a plant, just what fertilizer it needed and what its growing season was and thinking nothing of keeping her room at a certain temperature for days so that an Amaryllis, for instance, might have all the conditions favorable for blooming.

She filled her sprinkling can at the sink and let the tiny jets play on the box until the plants and the earth under them were drenched. Vesta would say that you ought not to water flowers in the morning, especially when the sun was hot as it was today, but she was likely to forget it unless she did it now, and besides, it was pleasant to see the earth turn dark under the water and have the leaves, freed from the gray film that collected on them every night, give off, if only for a few minutes, the warm, green smell of a real flower bed. She set her can on the sill and leaned farther out of the window. Earlier in the morning a sprinkler had passed by. The concrete was still steaming and in one place where the surface was uneven enough water had collected to provide a sparrow with a bath. He finished his bath and flew away, leaving a streak of sunlight, opalescent with oil, imprisoned on the bosom of the pool. She stared at it, her brow wrinkling. A phrase from her own thoughts came back to her. "An affair of this kind." "A hole and corner affair" was what Vesta had called it in that disagreeable conversation. "Is he going to marry you?" Vesta had suddenly asked. "You say he and his wife are separated. Is he going to marry you?"

She had been badgered into saying things she had never admitted even to herself. "I don't know," she had said. "I don't know that he wants to marry me. I don't know that I want to marry him."

"You aren't planning to marry him! You certainly are a fool, then, to go as far as you've gone. And you needn't tell me, Edith Ross, that you aren't living with him. I've got eyes in my head."

Vesta would be the one to give her away, blurting out some ill-considered statement before one of their friends, and then excusing herself on the ground that there was no use trying to hide things, that everybody knew all about it anyway. Yes, that would probably happen—unless they could get married soon!

She had hardly allowed herself to consider the possibility of marrying Chapman before. But, she asked herself now, why not? His wife had left him. He had told her some months ago that he had not been in love with Catherine for a long time, or she with him, and had said that they would probably be divorced. Catherine—she called her that in her thoughts now; she seemed somehow much nearer now that she had left him than when she was living with him—Catherine must want a divorce or she would not have gone away. And after all, she, Edith, was giving up a great deal for him, risking a great deal. He ought not to allow her to do that if he didn't want to marry her.

A breeze came in at the window, rustling some sprays of

green in a tall vase. Their elongated shadows, interlaced and quivering, rose from the floor and clambered up the west wall. She felt for a moment as if she were in a wood, with boughs stirring. It would be nice to be in the country today. There was a place in Pennsylvania where she and Chapman sometimes drove for dinner. A two-hundred-year-old stone farm-house, converted into an inn. You dined on a high, walled terrace, above a little stream, flower beds all about. The flag-stone path that meandered through the flower beds went on beside the stream and disappeared into a wood. She had contemplated the vista often without curiosity as she dined on the terrace, but now she found herself wondering what lay farther along the path. She tried to remember what her lover's engagements were likely to be for the day. If they drove out half an hour before lunch the inn people would make them some sandwiches and they could start out on that path and explore the valley. Or she might pack the lunch herself. There was some cold tongue and lettuce in the refrigerator and a good cheese . . .

She was investigating her stores when Chapman came in, fully dressed. She hastily took from the oven the toast she had kept warm for him, poured him a cup of coffee and another cup for herself and sat down opposite him.

"Don't you want an egg?" she asked.

He shook his head. He was unnaturally pale and the hollows around his lack-lustre eyes were shadowed. As she watched his lips grew tense, drops of sweat suddenly beaded his forehead. He had a hangover. He would not want to talk She sat silent, drinking her coffee and gazing at the floor where the long shadows of the privet leaves still quivered.

"What are you going to do today?" she asked after a moment.

"I don't know," he said.

"I thought we might go to the country."

He grimaced, holding out his empty cup. "If I get as far as Sixty-fifth Street I'll be lucky."

She got up and poured him another cup of coffee, then, coming up behind his chair, bent swiftly and laid her cheek against his. "Is it very bad?" she asked.

He did not answer until she was back in her seat, then he said, "It's all right, if I don't move around."

"You'd better go in and lie down," she said in a subdued tone from which the tenderness had suddenly ebbed. As she bent over him she had been struck by the rigidity of his pose. When she had laid her cheek against his he had not moved but had sat rock-like, as if he were only enduring the caress.

As if he read her thoughts he spoke, with a laugh. "I must have been pretty well organized when I landed here last night."

"Where were you?" she asked.

"Mike's. Bob Upchurch and Ed Ware were there. At least I think they were there." He laughed.

"You didn't answer when I called the apartment, around five," she said. "Weren't you there at all yesterday afternoon?"

"No," he said.

She got up and, moving to the window, straightened the green sprays in the tall crystal vase. Over her shoulder she said, "Jim, I've been wanting to have a talk with you."

There was a silence; then he said, "Anything special?"

"No . . . That is, nothing new."

She heard the legs of his chair grate on the floor and turned around, her hands thrust behind her, gripping the window sill. He was standing up, looking at his watch.

"I've got to be uptown in a few minutes," he said.

"I thought you didn't have any classes today."

"I don't. But I've got to see Reynolds. This is the third Saturday I've put him off."

"Don't you think it's time we talked things over?" she asked in a low voice.

He shot her a wary glance. "Yes," he said. "Yes, I do. But couldn't we do it later? I'm due up there at eleven-thirty."

"You don't have to go. You could put him off."

"I'd rather not," he said. He did not meet her eyes as he spoke and he stood stiffly, patient yet poised for flight.

"What are we going to do?" she asked harshly.

His eyes rested on her face intently for a moment and then he looked out on to the street. "You know my situation," he said.

"If you gave any thought to mine you wouldn't come up here at three o'clock in the morning and demand to be let in . . ."

"I was drunk," he said.

"That doesn't help me. That boy will tell the superintendent. I may have to move . . ."

"I know that," he said dully.

Something hopeless in his expression, his posture, aroused her compassion. She went to him quickly and, hiding her face against his body, put her arms up about his neck. "Don't let's quarrel," she whispered.

He stood motionless, then his arms came up slowly and

enclosed her. She felt their light pressure about her. It seemed nerveless, almost mechanical. She took a step backward. His arms fell away. He did not follow her.

He spoke. "I'll call you up. Perhaps we can have dinner together. I'll call you up. Right after lunch . . ."

The words rang in his own ears. Too smooth, too rapid, with something at once false and precise about them, they might have been uttered by a robot that he had conjured up, that he had disgorged, to stand there between them.

She stared at him, a faraway look on her face. She was trying to hear through the inhuman sounds other syllables. Suddenly she cried out and her hand went to her mouth and she turned away, her other hand plucking at the air.

Shadows of leaves stirred by the sudden movement ran in dapples up her white robe. Her shoulders bent, her hand still pressing her mouth, she stood, remote, lost in the airy thicket. A woman he would never call out to, a thicket he would not enter again. He spoke and afterwards could not remember what he had said before he closed the door and went on his brittle, metallic legs down the hall.

XI

ON THE TWENTY-SEVENTH of September Willy Lewis left home for the first long stay she had ever made in her life, taking with her the red stallion which she expected to show in the two-year-old class at the Fair. She was accompanied by Mr. Shannon, who had volunteered to show the horse for her.

That afternoon Catherine drove over to Big Pond, where she found Mrs. Manigault and Miller having tea on the shady terrace. Mrs. Manigault asked if Miss Lewis had left for the Fair.

"She got off," Catherine said. "My God, you should have seen the goings on! We were all up at the crack of dawn. Aunt Will's bag was packed. I thought she had nothing to do but get into the car with Mr. Shannon and drive off; then at the last minute she made a tour of the place. There were some pigs I'd never heard of. Six horrible little spotted pigs off in a pen in the woods. Uncle Joe is laid up and she doesn't trust Rodney, so she wanted Maria to promise to feed them herself. And Maria would keep saying how she certainly would like to feed them pigs, but she didn't see how she could get to them, with all she had to do. I *told* Aunt Willy that I would love to slop those pigs, but of course she thought I couldn't do it properly. Then at the last minute I got Maria off in a corner and told her I'd give her five dollars if she'd just *say* she would slop those pigs and let Rodney and me do it. She went into a tail-spin about how she was a member of the Corncob Baptist Church in good standing and couldn't tell no lie. 'All right,' I says, 'All right, Maria. I *was* going to town tomorrow and buy some things to send to Jesse. I was going to send him a big box of chocolate candy and some peanut butter and some sardines, but I'll be damned if I ever send him another thing, Christmas or no time, if you're going to act like this.' 'Gre't God, Miss Kit, you oughtn't to swear like that!' she said. But it brought her around."

"You were very clever," Roy Miller said. "Is this really the first time Miss Willy's been off the place this year?"

"She went to Carthage last spring, to see about the taxes," Catherine said. "But you can hardly call that a pleasure trip." She looked at Mrs. Manigault. "You know that dark gray shantung of mine? It fits her perfectly. She wore that, with a little red hat. She looked positively *chic*! Cousin Daphne was wonderful. She came in at the last minute and looked her up and down. I was trembling for fear she'd tell her the suit was too young for her and I didn't know *what* I could do with Cousin Daphne—knock her on the head, I suppose. God knows nobody'd miss her. But all she said was, 'Willy, I think I'd put on a little lip-stick.'"

"Doesn't your aunt own a lip-stick?" Miller asked curiously.

Catherine shook her head. "You don't know my aunt. She's really right out of Jane Eyre. Green eyes and all. . . ."

"And no Mr. Rochester has ever heaved over her horizon?"

"Never—as far as I know. That's the hell of it."

Miller, settling himself more comfortably among his green

and white pillows, gazed out over the lawn where a lone peacock was wandering and sang:

"Plaisir d'amour ne dure requ'un moment,
Chagrin d'amour dure toute la vie . . ."

He grinned, glancing at Catherine, and repeated the first line of the song. "*'Ne dure requ'un moment'* . . . Perhaps she's just as well off . . . God, Elsie, look at your bird!"

The peacock, as if aware that eyes were upon him, had suddenly whirled and stood facing them, his splendid tail feathers spread wide, his bronze eye implacably fixing a spot just to the left of the terrace.

"What do you suppose he sees?" Miller murmured. "Aha, it's love!" he ejaculated as the peacock, dragging his wings on the ground, advanced a few mincing steps towards a peahen that had come round the corner of the terrace.

Mrs. Manigault had smiled, looking at the peacock, but now her face wore the slight cloud that it always wore when poverty or any kind of necessity was mentioned.

"If Miss Lewis would get rid of those shiftless Robinson negroes and get one really efficient tenant, she could have an adequate income from that place," she said. "Tom says the river bottoms are the best in this country."

"But farming is so complicated," Catherine said. "You have to have the knack of selling as well as producing and Aunt Will completely lacks that. I don't think," she added reflectively, "that she could get along without the Robinsons. She *depends* on Uncle Joe. She even depends on Maria."

"Maria isn't the mammy type, I take it," Miller said.

Catherine laughed. "She's a good mother to her own children. Mammy always says that she's a good neighbor—but as for service! When you get up from one of Maria's meals you always feel that you're lucky not to have been poisoned. She serves things so *darkly*. Once she told me that if people only had the moral courage to quit putting food into their stomachs the Lord would solve all problems by taking them away from here."

"A colored Schopenhauer," Miller said. "Catherine, I do love to hear you talk about Swan Quarter. You're the only pretty girl I know who can talk just like a book."

Catherine eyed him coldly. "It comes from associating with Jim," she said.

"Is he coming down soon?"

Tom Manigault, in khaki-colored trousers and shirt, carrying a metal part of some machine in his hand, came out of the house on to the terrace. Catherine greeted him before she answered Miller.

"In a week or two," she said, "if he can get away. I'll have to stay here now till Aunt Will gets back."

Tom sat down, laying the round, cogged wheel on the floor beside his chair. He took a handkerchief from his pocket, wiped a spot of grease from his thumb and accepted the glass of iced tea his mother offered him.

"Miss Willy got off?" he asked.

"She and Mr. Shannon drove away at a quarter past seven," Catherine said. "Tom, she asked me to thank you all again for that horse van. I think she might have backed down at the last minute if she hadn't seen how luxuriously Red would travel. Mr. Shannon was ecstatic. You know, he's awfully sweet. You'd think Red was his own horse. . . ."

"Marble is rather worried over Red's prospects," Mrs. Manigault said. "He thinks Miss Willy might have done better to wait and show him in the three-year-old class."

Tom laughed. "He's worried over that colt of ours. He hates to see a woman breeder get ahead of him."

"You can hardly say that Miss Lewis bred Red," Mrs. Manigault said dryly. "He was dropped right here in one of our stalls. I remember the night."

"It was when I had pneumonia," Tom said. "You loused things up. Told that old nigger you didn't have time to fool with the colt and tried to get him to kill him. If you'd kept your eye on the colt instead of me you'd have a winner and you might have got rid of some of your troubles too."

"Are you implying that I would be happier if I had allowed you to die of pneumonia?" Mrs. Manigault asked.

"You called the turn," Tom said cheerfully, "I didn't."

Catherine watched their faces uneasily. She had never been fond of Cousin Elsie. She was too energetic, too dominating, too *moneyed* for her taste, but of late there was something about her that aroused compassion. She was, after all, getting on in years. She did not stand up well under the attacks that Tom made on her. Sometimes when he had said something particularly shocking she would lower her handsome white head and glance furtively out of the corner of her eye as if she hoped nobody else had heard him. A moment ago she had looked over at Roy as if she expected him to come to her assistance. And indeed he usually took her side when Tom quarrelled with her. But today he seemed preoccupied, his head half turned away from the company, following with his eyes the movements of a slim, young colored boy who was playing a hose on a brown spot in the lawn. His thoughts, whatever they were, seemed pleasant. He was humming under his breath:

". . . coulera doucement
vers ce ruisseau qui bord de la prairie. . . ."

"Now what's the matter with *him*?" she thought.

She stood up. "I must get back to my responsibilities. Tom, how much corn would you give four-month-old pigs at their night feeding? Rodney and I disagree."

"Three or four nubbins apiece," he said absently, and then in a sharper tone: "Aren't those niggers looking after the hogs?"

"I'm overseeing the job. I don't trust Rodney."

"I'll feed them for you," he said.

She shook her head. "No, you won't," she said in a tone that she made more decisive because of the sudden, inquisitive glance that she fancied Miller bent on her. "Indeed, you won't," she repeated. "These are my pigs."

She bent and kissed Mrs. Manigault, kept Miller in his seat with a swift, levelling motion of the hand and turned to the door, but Tom was there before her. "Want to give me a lift down the road?" he asked. "I've got some men working in the north field."

"Yes," she said.

When they were outside, in the car, she asked: "Why did you do that?"

"Ask you for a lift?"

She shook her head impatiently. "No. Say you'd come over and feed the pigs."

"Don't you believe in being neighborly?" he asked.

She paid no attention to what he had said, pursuing her own train of thought. "They know you never do anything when you can get a nigger to do it for you. They'll think you're crazy."

"They *know* I'm crazy," he muttered.

"Oh, Tom!" she exclaimed, and drove more rapidly down the road. When they came opposite a group of men operating a tractor she halted the car. He did not move. She did not seem to notice that he was not getting out, absently following with her eyes the movements of the men and the machine. "Mr. Miller seems in a curious state," she said at length.

"He always is."

"I know . . . But lately . . . Oh, I don't know. He's always talking about love."

"He's been going on about love ever since I can remember." He laughed. "Reminds me of an old mule we've got. Nothing Henry can do about it but let a mare come around, and Henry's pawing the ground and hollering loud as any stallion."

"Poor thing," she said abstractedly. "I suppose it's on his mind a lot."

"Who? Henry?"

"Roy, you idiot. Well, there's nothing he can do about it—down here, at any rate. . . . Tom, do you suppose there's a person like him in the whole of Carthage?"

"Sure," he said easily. "Didn't you ever see old Mr. Ralph Bateman? Wears that long black cloak down to his heels and sticks partridge feathers in his hat. They tell me that even some of the niggers are that way these days. I'd draw the line on that. I wouldn't have a nigger on my place if I knew he was a fairy."

"That's absurd," she said lightly. "They've got just as much right to vice as we have."

"I know. But there's something about it puts my back up. I feel that if a white man's that way it's his own business. But a nigger is different."

She turned her head, regarding him with an amused light in her eye. "Tom, you really are wonderful. What I can't understand is how you got that way, considering your bringing up."

"I never had much bringing up," he said. "One summer, when I failed in Latin, they got me a fairy for a tutor. Mother was in Europe that year and Dad was staying in town nights, so Edward and I had the place to ourselves."

"*Tom!* He didn't try to corrupt you?"

He shook his head. "He was a pretty good guy. We'd bone all day—he worked the stuffing out of me—and then he'd have a party at night. There was one fellow brought a trunk every time he came out. Used to change his dress two or three times during the evening. Wore a kind of gold wreath on his head one night, and bracelets."

"You mean you were allowed to witness these orgies?"

"Well, in a way. The butler wised me up and I used to pretend I couldn't sleep and go down and get him to make me some hot chocolate, and then I'd stick around in the pantry and we'd take a look at them every now and then . . . The whole bunch got canned when one of 'em dived into the swimming pool and cracked his skull. It had been emptied, you see, but he didn't know it and took off like a swan, just before day-break one morning. It was hard on Edward. The fellow had a fractured skull and sued us for twenty thousand dollars. And of course we maintained that he didn't have any business in the pool, anyway, and Edward countered that his contract permitted him to entertain friends on the grounds. Made quite a stink."

"I should think so. Did you have to pay?"

"I don't remember. Mother probably wrote him a cheque. They threw money around like water in those days."

"What a horrible thing to happen. Well, anyhow, they didn't corrupt you."

"No," he said, his bold eyes roving her face. "No, I'm not very corrupt."

She laid her hands on the wheel. "I've got to go. Aren't you getting out here?"

"I'll go on with you and walk back," he said.

She looked at him steadily. "Do you think that's a good idea?"

"Why not?"

She shrugged her shouders and started the motor. Neither of them spoke until they had turned off the highway into the lane that ran between Swan Quarter and the Shannon place, when he asked, "Is Chapman coming down?"

"I don't know whether he is or not . . . I made that up a minute ago. Mr. Miller is so damned curious."

"You've heard from him?"

"I had a letter last week."

"What did he say?"

"He said that it was very hot in New York . . . Oh, Tom, I don't want to talk about it."

"I know I haven't any business asking you questions," he said.

She laid her hand on his arm. "Tom, I'm so fond of you. I wish you hadn't taken this notion into your head."

"You mean wanting to marry you?"

"You *can't* marry me," she said impatiently. "I'm already married."

"I don't expect to marry you next week or next month," he said, "but it's pretty plain that you and Chapman are on the rocks. You are, aren't you?"

"Yes," she said in a low voice, "but I don't want to talk about it."

"I won't talk about it," he said. He took hold of the wheel and began turning the car off on to a wagon road that led through the woods.

She shook her head, smiling and keeping hold of the wheel. "Tom, I was awfully upset last week. I hardly knew what I was doing. I was very unfair to you, very inconsiderate . . ."

He squared around so that he could look down into her eyes. He laughed. "You mean letting me make love to you? I don't think you were inconsiderate. I think you were sweet." He suddenly slid both arms around her and gripped the wheel.

She resisted for a moment, then let him guide the car off the lane into the narrow, rutted road. When they were out of sight from the lane he halted the car and quickly got out. As she descended he caught her in his arms and held her to him for a moment, murmuring her name.

She slipped away from him, flushed and shaking her head.

In silence they left the road and took the path up to the spring. Halfway up the slope she stopped and sat down on a rock. He remained standing beside her, looking down into the ravine.

"Miss Willy ought to sell some of that white oak," he said. He pointed to an oak whose boughs towered above the rest. "That tree's ready to be cut now. In a year or two it'll start rotting."

"I don't believe she'd do it," she said. "She has a horror of cutting down the woods. They were always rowing over that when Uncle Jack was alive."

"She doesn't need to cut down the woods," he said, "just take out the trees that need cutting. There must be a dozen of those white oaks in here. Joe Applegate'd move in here for that much."

"You mean set up a saw-mill in here?"

"Sure . . . But I wouldn't trust Joe to pick the trees. And she'd have to watch him while the mill was running."

"She couldn't do that. And she wouldn't have any idea which trees ought to be cut."

"I could go through the woods and mark them for her. And I could keep an eye on Joe too."

"Would you do that?" she asked eagerly. "Would you do that for her, Tom?"

"Sure," he said. He turned around, grinning. "You don't know what a good neighbor I am."

She smiled at him, thinking that it was true. He had a really good heart. "You like to live down here, don't you?" she asked.

"I wouldn't live anywhere else now," he said. "You know,

there's a stretch in November, after the leaves have turned when it's mighty fine down here. You get up early in the morning and ride out over the place . . . There's going to be a lot of quail this fall, Kit. I've got a little gun that'll just suit you . . ."

She shook her head. "I can't bear to shoot things."

"You'll get over that."

She stood up suddenly. "Look!" she said and pointed to something shining at the foot of the path.

"Did you drop something?" he asked. "I'll get it."

But she was already at the foot of the path, stooping. She held the object up for him to see: a round silver box with chased edges. She thrust her hand into the pocket of her jacket; there was a slit in the lining. "It's a wonder I ever found it," she said.

They climbed the slope again and sat down on the rock. They sat in silence. Presently he took the box from her, lifted the lid and, finding that it contained nothing but a few grains of powder and a puff, handed it back to her. "I thought you must have had some money in it," he said.

"No," she murmured, "just my complexion." She continued to hold the box in her hand, turning it this way and that so that the rounded surface caught the light. The decorative border was darker than the rest of the lid. It's dirty, she thought, I must polish it, and she bent her head to examine the chasings. She had had the box three years but until this moment she had not realized that the design was sheaves of wheat interspersed with some round fruit. I never liked it, she thought. That's why I never really looked at it before.

Jim had bought it for her on her thirty-second birthday. When he gave it to her she had been surprised, for he was not the kind of man who made presents. Once, before they were married, walking on Madison Avenue, they had passed a florist's shop and he had broken from her side with a grunt and disappeared into the shop while she lingered on the outside, looking at an unattractive bouquet of red roses and yellow snapdragons. A huge sheaf of the same roses and snapdragons had been delivered to her a few hours later. But he must never have passed a florist shop again during their courtship, for no more flowers had arrived, and she could not remember any other present he had ever made her. The compact had been Molly Ware's idea. It would have been better if she had gone with him to select it. They had sold him this preposterous little number that was all looks and had no room in it for either lipstick or rouge.

Tom Manigault stood up. "I'd better be getting back," he said. She stood up too. He did not go immediately but stood before her, not saying anything, but keeping his eyes fastened on her face. His mute look said that there was no necessity for his leaving, no concern of his own called him from her side. It was merely that he found it too disturbing to sit here beside her. He must have surcease for a time from his own violent emotions. A love affair, she thought, if it got anywhere at all, always reached that stage with such alarming rapidity.

She stepped past him. Her hand, going down into her pocket, touched the little silver box. "I could fling it out now," she thought, "fling it out and let it tarnish here under the leaves." Her hand closed on the box. She started down the path. Tom followed. He had accepted her rebuff quietly but tomorrow or the next day he would renew the attack. Three weeks ago he had begun to make love to her. Sometimes she rebuked him as she had just done. Sometimes she listened to the pictures of the life they could have together, if, as she sometimes put it to him, laughing, she would only start for Reno in the next half hour. His mother could keep the big house, he said, and he would build a house for them on the other end of the farm. She would enjoy living in the country. Would she like to raise horses? He knew where he could lay his hands on a good brood mare. A High Cloud colt. If she would say the word he would buy the mare now. She could breed her before she went to Reno. Have a colt waiting for her when she got back. "Shut up, Tom," she had said, "you're indecent," but she had listened to him and would listen again. There was a fascination in it, like getting steamship folders and planning the details of a trip you knew you would never make. And there was relief, too. If there came a time when you could not bear this life that you had, it seemed that you might escape to another life.

XII

AT HALF PAST SEVEN Maria entered the cabin. Old Joe was lying flat on his back. He raised himself with a groan and, doubling up one of his pillows, propped himself against it.

"You needn't be settin' up so high for what you got," Maria said.

He sighed contentedly. "I reckon I got something," he said. "Ain't never been a night they didn't send me something."

"War'n't much supper tonight," Maria told him. "Least nothing you'd eat. But I got you some snaps here. And Miss Kit sont you some cheese."

She went to the iron kettle which sat on the hearth and ladled some snap beans and a strip of salt pork on to the plate she had brought from the house. Joe rested the plate on his hunched-up knees. He took a bite of the cheese before he ate any of the pork or beans. "Jesse goin' to relish this," he said. "She got the box off, didn't she?"

"Was some of that cheese," Maria said, "and cakes; what they call Fig Newtons. And some chocolate cakes, too, with icing. And a box of candy. She didn't send no sardines."

She pulled a rocker up to the hearth and sat down. The fire that she kept burning there at all seasons of the year smouldered dimly through ashes. She drew the ashes away with a poker and, taking up a turkey wing fanned the embers until they glowed, then laid some chips on top of them and, leaning back in her chair, stared into the rising flames.

Joe had finished his supper. "Here, woman!" he called sharply. She rose and took the plate and knife and fork from him, washed and dried them and resumed her seat.

He remained propped on his pillows, surveying the room, lit now by the dancing flames. He ran his tongue over his lips, then wiped them on the back of his hand. "I reckon she'll send some sardines next time," he said.

"May not be no next time," Maria said.

Joe was silent. "You slop them hogs?" he asked suddenly.

"Miss Kit's tending to them hogs," Maria said, "her and Rodney."

Joe laughed. "Miss Kit's just like Miss Agnes. Likes to be around stock. Miss Agnes used to go with me to feed the stock when she was so little she couldn't hardly get through the grass. And after she's grown she used to pick her out a calf every year and shell corn for it herself . . . Got out in the lot once when I was butchering. Had my maul raised to fell a steer and here comes Miss Agnes. 'I'll kill you, Uncle Joe!' she says and she bit me on the hand till I had to drop my maul. I told Mister Jack, I says, 'I can butcher anything on this place you want butchered, but how anybody going to butcher with Miss Agnes swarming up 'em like a weasel?' Mister Jack, he told me to put away my maul, said we wouldn't butcher nothing else that day. And he sont that steer off the place."

"It was part your steer," Maria said. "He didn't have no right to do that."

"He paid me for it," Joe said. "One thing about Mister Jack, he war'n't chinchy."

"He war'n't chinchy," Maria said, "but he was drunk

more'n half the time. White folks take a drink, it ain't nothing. But let a nigger get one too many . . ." She sighed. "Well, he's dead now, and underground."

"I wish he war'n't," Joe said. "I could always git along with Mister Jack."

"You could always git along with any of 'em."

"It ain't hard to git along with Old Miss," Joe said. "All you got to do is git her in the lead and keep her there. Trouble with you, you want to stay in the lead yourself."

Maria did not answer.

"Fill me my pipe," Joe said.

She took a hand of tobacco down from the mantel and, pulling off part of a leaf, crumbled it in her hand, filled the bowl of the corn-cob pipe, then crouching on the hearth, dexterously picked up a live coal and lit the pipe and drew on it a few seconds before she took it to him.

When she handed him the pipe she looked down at the pillows wadded behind his back. The slips were dingy. They should have been changed this morning. But she would not disturb him now. Her eyes sought the far corner of the room where the bolster lay across a trunk. The sham which covered it was ornamented with huge purple morning glories and trailing green vines. She remembered the day she finished working those morning glories. She could hardly wait until she got the bed made up, with the blue-and-white quilt, the huge pillows and finally, on top of all, the bolster. The morning glories had stood out as bright as real flowers, so pretty with the green leaves all around them. She had left her work two or three times that day to run back to the cabin to have a look at them. That thread hadn't faded in all the washings it had had; the colors of the morning glories were just as bright, the cloth itself still spotless, but she took no pleasure in the sham now, would just as soon have had any old rag on the bed.

She went back to the fire. The hand of tobacco was still lying on the table. She broke off half of a leaf and, stuffing it into her mouth, chewed slowly.

"Miss Kit say anything about leaving?" Joe asked.

She shook her head. "She don't say nothing to me and I don't say nothing much to her . . . She's carrying on with Mister Tom Manigault. Least I see 'em meeting off in the woods."

"Maria!" Joe said with sudden energy. "There's some of the folks over at the church wouldn't think as much of you as they do if they had to live with you. You know that when white folks go off walking in the woods it don't mean the same as it does with niggers."

Maria was silent, chewing and looking into the flames. The nicotine was having its effect. Her mind was as clear as a bell now, but in a few minutes it would be even clearer. In a few minutes it would be like she wasn't in this room.

She glanced at Joe, still propped on his pillows, smoking quietly. Standing there beside him, she had looked down and had seen the top of his head all white. It had not been like that last year. Those groans and grunts weren't all for his rheumatism. He was breaking fast. He had had enough to break him. In a few years he would be past work and they would have to go to town, to live with Elvira and her husband. Joe kept thinking that Miss Willy would get on her feet, that tobacco would get high again and it would be like it was in the old days when folks could afford to take care of their hands when they got past work. But things never would be that way again, and even if they were, Miss Willy wouldn't catch up. There was hardly enough for her and the old lady now. She couldn't take care of two old people who were past work. They would have to go to town. She herself didn't care now where they lived. But it would be hard on Joe, to live where he couldn't even have a hog or a garden spot, and to have to depend on Elvira. They had never gotten along, and as for Elvira's husband nobody could get along with him. The roistering kind who'd sell the legs out from under the table to get a drink . . . She wished that she hadn't said that about Miss Kit meeting Mr. Manigault off in the woods. There was no use in telling everything you knew, especially if you knew more than other people.

She tilted her head slightly, gazing through half closed lids. The sycamore tree just outside the door stood up lightly, its branches spread like arms. If she turned her head now and looked behind her the room would seem different from the way it had a moment ago. Everything would stand out clean and light, as if it could move around by itself. That was always the sign, when things got that airy look. It didn't happen every time she chewed tobacco. In between times she would forget, and then all of a sudden, she would get the sign and know that for a few minutes, anyhow, she would have a little ease, the only ease she ever got. She could bear to think about anything then.

It was strange about Elvira's husband. He had been drunk off and on since he was fourteen years old, hardly ever did a day's work and was roguish to boot. And yet he had a good name and if he were willing to leave the crap games in Carthage could walk out this Christmas to any farm in the county and make a good trade, while Jesse, who was a good worker and had never touched a drop till he was past eighteen, had not been able to get work anywhere in the neighborhood when he got out of the pen.

That was how he had come to shoot that other man. Working off on a strange place, twenty miles from home, among strange hands, he had been scared and had acted proud to cover up. Roscoe Davis, whom everybody thought so much of now that he was dead, had been in more fights than Jesse had. It was he who had started the trouble, drawing a gun and coming at Jesse. No wonder Jesse had got out his razor. It would have been better if Jesse had had a gun on him too. A razor moves faster than any trigger and has a will of its own: a man can't tell how deep he is cutting. The razor had sliced through into Roscoe's left breast . . . And now Jesse was behind walls. For the rest of his life. Her lips trembled. "I'm with you, son," she whispered, "I'm in there with you . . ." But it was for life!

That was what Mister Jack had said, standing there in his room that day, his face all white with lather.

"I tell you he's in for life this time, Maria!"

"Couldn't you write to the governor? Last time you got on the train and went up there to see him."

He scraped some of the lather off his face and flung it down hard in the basin.

"I got him off last time on good behavior. He was on parole to me and he hadn't been home three months before he killed another man."

"They tell me that that Roscoe was mighty bad about slipping up on people and cutting 'em," she said.

"He was a bad nigger, all right. But he hadn't ever been in the pen . . . Jesse ought not to have gone to that barbecue. And he ought not to have had a razor on him."

"He war'n't nothing but a boy, Mister Jack. Boys all carry razors."

He did not answer, turning around to the washstand, taking his razor up and guiding it over his chin. She watched it flashing in the sun and for a second it was as if her own hand had hold of it. She would smooth off the chin just as he was doing and then, suddenly, slip down on the neck and cut in deep, cut in, even after the blood had started spurting.

She felt sick at her stomach and went downstairs as fast as she could.

The old lady had heard her passing through the hall and made her come into her room. "I knew you were wasting your breath, Maria. Jack told me there wasn't any chance, after Jesse killed that second man." She was darning and she let her darning gourd roll off her lap on to the floor and Maria had had to stoop and pick it up for her. "And besides, if we did get him out, you'd never have a minute's peace. It'd be no time at all before he'd be in trouble again . . . You spoiled that boy, Maria. Why, he was eleven years old before he learned to plow!"

She had got away. There was a chicken waiting for her, in a pot, on the back porch. She sat there all afternoon, picking it. A Black Minorca. It takes a sharp hand to pick a black chicken clean. But she was sorry when the feathers gave out.

It was the next afternoon that Mister Jack had been killed, his neck broken when he fell from a wild horse. She was in the cabin, resting, when Joe came running. "His *neck?*" she cried. "It ain't his *neck!*" But the doctor was there by that time. He said that his neck was broken the minute he hit the ground, quicker even than a razor could have cut it . . . She did not tell Joe that she had wished Mister Jack's death on him. He was nearly out of his head as it was. But she never had anything in mind but to confess, to stand up in church before everybody and confess. But meeting didn't start at Concord for two weeks. And by that time she had got used to the thought of her sin. She had never confessed, had never told anybody; she had just got used to it.

There were times when she wouldn't want to give it up. It made her feel closer to Jesse. They had both killed, only she had meant to, and he hadn't. She knew he hadn't. It wasn't in him.

She allowed herself to recall Jesse as a little boy. He had had typhoid fever when he was seven years old. That was why he was so late getting into the field. She had had to keep him around the house longer than the other children. And he thought more of her than the other children did, seemed somehow to have more love in him. That year that Miss Kit had had a woman from Big Pond cooking for her she and Jesse had been here in the cabin together all day long. He liked to sort quilt scraps and sometimes he would twist them into pretty flowers. When she fetched water from the spring he went with her and, coming back, would bring as full a bucket as she would let him. "What you going to do with all this water, Mammy?" . . . "Wash the white folks' clothes." He stood there on the path and looked down at the house. "When I get big I ain't going to let you wash their clothes for 'em." . . . If he hadn't taken to playing the banjo it would not have happened. He could play better than any of them and they made him play for the dances. Somebody always brought a jug. Some men could drink whiskey and lie down and sleep it off. But Jesse was high strung. Let him get a drink in him and he didn't know what he was doing . . . Ought a man to be punished when he didn't even know he was doing wrong?

She drew a long, shuddering breath, then glanced behind her in alarm. But Joe was sleeping, prone under the covers. She got up and went to the door. Mist was rising from the creek. The lights from the house showed through it dimly. When she left the house they were all sitting on the porch, but now lights showed in young Miss Kit's and Miss Daphne's rooms. The downstairs bedroom was dark. Old Miss must be in bed. Young Miss Kit was trying to take care of her while Miss Willy was gone. No matter how hard she tried she didn't please the old lady. She was used to Miss Willy, who would lie down and let her walk over her a dozen times a day. But no matter how much Miss Willy did for her the old lady would never think as much of her as she had of Mister Jack.

If it hadn't been for old Miss Kit she would have confessed and been rid of her sin. She had sent for her right after Mister Jack was killed. She didn't want to go, but Joe made her. "Go on, woman. Go on. You *got* to go!" . . . Miss Kit was in her room, by herself, trying to take off her clothes, but her fingers wouldn't serve her and her legs gave way and she almost fell, stumbling to the bed. "I feel so weak, Maria . . ." "You be better in a minute . . ." She had covered her up and was leaving the room when the cry rang out. Not a woman but a wild thing, roaming the woods and shrieking its grief into the night. Kinfolks had come to look after Miss Willy. She had stayed with the old lady, steeping cloths in water and vinegar and laying them on her forehead. Miss Kit would let the cloth stay on for awhile and then it would come over her again that he was dead and she would rise up, moaning. Towards morning she fell asleep. She herself was dozing in her chair when the old lady woke her, saying in a quiet voice, "Maria, has anything happened to Jack?"

"He's dead," she said, "killed, falling off his horse."

Miss Kit had got up out of bed and started across the room. Her nightgown was open down to her waist. Her breasts were showing. But she didn't care, running to the window and then to the door and all the time shrieking. They came and gave her some medicine the doctor had left and the shrieking stopped, like a chicken's when your hand closes over its throat, before you wring its neck. But you could still hear it in the room. And all the next day out in the cabin she could hear it. She called it up now sometimes when she wanted ease. She stared up at the watery sky. Mister Jack was dead, and buried over there on the hill. And Jesse was behind bars. She would never confess while Miss Kit had breath in her body. I'll keep her crying out, she said, crying out for both of them.

XIII

AFTER SUPPER they sat on the porch, but at half past seven the old lady rose to go to her room. Catherine went with her to help her undress. Daphne hesitated, looking out over the lawn, where fire-flies were already flitting. "I think I'll take a turn in the woods," she said. "It won't be dark for an hour yet."

Catherine said, "Cousin Daphne, mushrooms must be the most absorbing of all passions. You never give up hope!"

Daphne shook her head. "I've just thought of a place where there might be some *Russulae*. I've about got time to make it before dark."

She went upstairs and got her basket and knife and set out for the woods. Catherine was right, she thought. Of all the hobbies she had ever had mycology was the most rewarding—it was a hobby, of course, not a passion; Catherine had an exaggerated way of speaking—but it was better than Italian cut-work or burning designs in leather as she used to do in her young days, better, even, than knitting. One never knew what a day would bring. If one were indefatigable and resourceful, as she was, one need never give up hope. Now for a week she had roamed the woods, finding nothing but one good-sized "Bear's Head," growing out of the trunk of a tree, and then, a minute ago she had thought of a place that stayed damp in the driest weather: the ravine north of the graveyard. A little stream ran through it. Its banks were covered with moss: the kind of place that *Russulae loved.*

She arrived at the edge of the woods, hesitated, then struck north. A few minutes' walk brought her to the bed of the ravine. The stream was low, hardly more than a trickle of water, but the mossy banks were spongy with moisture and even the dead leaves on the slope were damp. A path ran beside the stream. She walked along it slowly, bending a sharp eye on the moss. As a child she had come here, with Agnes Lewis—they had had a playhouse in a cave up the slope and dug moss from beside the stream to decorate it. There had been mushrooms growing here then, tiny, scarlet trumpets nestling in the moss, stouter mushrooms of a deeper scarlet, thrusting up through the dead leaves. But they had believed that all mushrooms were poison and had paid no attention to them, except occasionally to stamp one underfoot.

She remembered one mushroom that they had found, standing six or seven inches high on its slender, hollow stalk. They had admired its pale brown top, ornamented with deeper brown spots and had thought that the under-side looked exactly like cream-colored, accordion-plaited silk. Agnes, in spite of all she could say about its being poisonous, had carried the mushroom up to the cave on the hill, to be her lady's "parasol." The *Lepiota procera* or Parasol mushroom. What would Agnes say if she knew that she was hunting a mess now for her dinner?

She found a lettuce-like *Sparassis crispa* growing beside a tree trunk, desiccated but a splendid specimen. A little farther on was a cluster of green *Russulae*. Some of them were showing signs of decay but four were in their prime, firm and meaty, each as big around as a coffee cup. She plucked them and as she straightened up glanced at the sky, showing gray through the interlacing twigs. She ought to be turning back. It would be dark very soon. But if she went a little farther she might come on more *Russulae*. It would be a satisfaction to have enough for a mess. They had eaten the last of the Bear's Head at supper.

Ahead the trunk of a sycamore glimmered. There was a little pool between its arching roots. It was there that they used to come to wash their "dishes," pieces of broken china or glass that they had brought from the house, or bright-colored pebbles that they found in the bed of the stream. In those days there had been a path going up to the cave but it was obliterated now.

She stood beside the sycamore, staring down into the dark water. She remembered the day that Agnes had said that they would have a playhouse and had led her up the slope to the cave. It was a rock-house, really, for there was no passage into the earth. But the overhanging slab was as wide as the roof of Maria's cabin, the rocks that formed the walls were half as high, and the floor was covered with fine, white sand. "We'll put a rock wall down the middle," Agnes said, "and you can have half and I'll have half."

They had toiled all day, erecting the rock wall, and then had set about devising furniture for the house: benches made of flat rocks, supported on smaller rocks, wider slabs of rock, covered with moss "mattresses" for beds, little moss rugs scattered over the white floor. There was blue clay on the banks of the stream. They had made their "people" out of that. Small, shining pebbles for eyes, moss garments, cheeks colored red with pokeberry juice. If you sprinkled the figures lightly with water the clay women—they had not bothered to make any men—would not crumble but stayed intact for days.

She set her basket down and started up the slope. In one place some trees had been felled and a kudzu vine sprawled over the stumps in a great tangle. Beyond the vines shadows were massed black on the side of the hill. She skirted the tangle, climbed over a heap of rocks and stood in front of the cave. It was smaller than she remembered it and the shadows were so deep that she could hardly tell the rocks from the bushes. But the sand that still covered the bed of the cave gleamed white. There was a dark line running down its middle. She stooped and felt stone cold under her hand. Those old rocks that had made their dividing line were still there. She let her hand brush over them, feeling the edges rough under her fingers, and turned and stumbled down the hill.

She had set her basket against the trunk of the sycamore or she might not have found it, darkness was falling so fast. She hurried along the path, one arm crooked in front of her, protecting her face from the bushes. Ahead the path was white; mist was rising from the branch. She had not been this way in thirty years. Not since that summer. It must have been the summer that she was twelve and Agnes nearly fifteen. Agnes had been a strange girl, large for her age, quiet among the grown people. She would sit in the house and sew and talk with them as if she were grown, too, and then suddenly slip back into childhood, have a spell of playing paper dolls or make up some Indian game. She herself still had all those paper dolls they used to play with. The Bascomb girls' paper dolls were all named Mrs. Bascomb or Mrs. Lewis or Mrs. Manigault, but Agnes would take some pretty, rosy cheeked creature out of *The Designer* and turn her into Fair Rosamond or Eleanor of Aquitaine. In those days, all the children visited at Swan Quarter in the summers; she, Daphne, had been Agnes' favorite. That was because she had imagination. Agnes knew that if she thought of anything interesting to do she would follow her while the Bascomb girls were wondering whether they would get their feet wet. There had been one summer when she and Agnes divided the woods between them. They were Indian chiefs, who met at the "Peace Tree," surrounded by their retinues . . . But the rock-house had been the best secret they ever shared. There was something mysterious about the place that brought them back to it day after day. There was always something new to do: moss mattresses to arrange on the beds, the clay people's cheeks to be colored. They had played there a week and then she had had to go into Carthage to visit—she and her mother used to leave their Nashville boarding house in June and spend the summer making the rounds of the relatives—and had spent two weeks at Uncle Will Morton's; there were no children, and the yard was no bigger than a pocket-handkerchief. All the time she had been longing for Swan Quarter. And then one afternoon the Bascombs had come by in their surrey and they had all driven out to Swan Quarter.

There was company there before them: the Beekmans from the next farm. Water-melons had been brought out and were being eaten on the lawn. She had whispered to Agnes that maybe they could give the Bascombs the slip—

she wanted to go to the rock-house—but Agnes only smiled uncertainly and a minute later when Tom Beekman came up she moved with him to the end of the lawn. The other young people followed them—there was a rope swing hung from the limb of a big oak. She had not gone with them, even when Cousin Kit came up behind her and gave her a push, but had sat there on the red bench, and watched Agnes get into the swing, knowing, even before Agnes soared up into the air, laughing, her skirts fluttering about her legs, that it was all over and Agnes gone from her, into another world.

Afterwards she had had moments of rebellion, moments, even of hope. She had thought that Agnes would get tired of having beaux and come back to her. There had been nights when she lay awake, praying that she would. But deep inside she had known that it would not happen, that Agnes was gone, never to return. They said that childish griefs were soon cured. She could not ever remember having experienced a greater sense of desolation. And why not, she asked herself, bringing her arm up savagely to brush a cobweb from her face. Why not? It was the last time in my life that I was ever happy.

She emerged from the wood and came in sight of the house. The lower part yawned as black as the cave on the hill. The old lady was evidently in bed. But a light burned in Catherine's room. If Willy had been there a lamp would have been left burning on the table in the front hall, but Catherine, of course, would never think of that. She crossed the porch, stumbling against a rocker, slipped through the door, locking it behind her, and went up the stairs.

Light, showing in a slit under Catherine's door, cast a faint glow into her own room. She made her way to the bed-side table found a match and lighted a lamp, then closing the door, she turned the key in the lock.

When she had undressed and had put her red flannelette robe on over her nightgown, she laid her specimens out on the table. The green *Russulae* she placed on the window sill. They ought to have gone into the ice-box but she did not feel like going downstairs again and it was cool enough in here. They would do just as well on the sill. She opened her note-book—its gray cover was ornamented with neat red lettering: "An Encheiridion of Wood and Field Mushrooms, by Daphne Passavant"—numbered a page and wrote, *"Sparassis crispa* (*Herbstii*)."

She examined the *Crispa* through her magnifying glass. Even when its petals were desiccated they retained their original texture, gelatinous and at the same time firm. The color did not change, either. But how to describe that color, which was neither gray, nor green, nor yellow, but some unnamed color, a color a plant might develop growing in an underground cavern, or in another world. For a year she had tried unavailingly to find the word for that smooth, gray-green-yellow.

She remembered the first mushroom she had ever plucked. She had been walking in Centennial Park in Nashville, with Mrs. Mayhew. They had walked up over a rise and suddenly a yellow flame had seemed to sweep over the grass: half a dozen clusters of *Clitocybe illudens,* growing at the foot of an oak.

"Look at those things!" Charlotte had said and they had

gone over and stood looking down at them. They were as large as calla lilies, pure yellow, shaded with scarlet. If you got an *Illudens* at a certain stage and put it in a dark room, it glowed. And in the daylight that incandescence seemed to linger in the petals. They were so beautiful that she stooped and picked a cluster. Mrs. Mayhew touched one with her finger, then took it away quickly. "It feels clammy," she said. "It feels like dead people." And from that day forward she would never admit that there was any beauty in a mushroom.

Charlotte Mayhew had been dead herself for ten years. She never thought of her as dead, but as sitting in the blue shell chair in the sitting room of the boarding house on West End Avenue. It was the only really comfortable chair in the room but nobody else would have thought of taking it. It was always left free until Charlotte, having given her orders for the next day, and having conferred with Uncle Will—she never could have run that boarding house without that old negro man—swept into the living room where the boarders who had nothing else to do were congregated. She would have stopped in her bedroom to touch up her hair and face—sometimes she even changed her dress—but she never joined old Mrs. Trabue and Mrs. Sloane and Mrs. Sawyer at their bridge, but would sit and look on. Sometimes she would roam about the room, straightening a picture or looking out of the window or turning the radio on too loud. If you were playing cards you might glance up and Charlotte would have been standing there several minutes looking at you. She would be smiling, as if there were something innately ridiculous about you and she would not change her expression but would broaden her smile, as if inviting you to laugh at yourself. Or she would tap a card with a red, pointed nail. *Play the queen, Daphne! Play the queen and see what happens,* and then you would have to try to finesse your queen, even though you had decided that it couldn't be done.

People who had known Charlotte all her life said that when she was a young girl she had not had that streak of malice in her nature. No doubt her hard life had brought it out in her. It is no joke to be left a widow, penniless, and with three children. She had no education, no particular talents. There was nothing for her to do but open a boarding house. But she had never made any secret of her hatred of it. Old Mrs. Trabue had told her, Daphne, that that was why she was leaving. She couldn't bear sitting at a table where the hostess was so indifferently amiable; she was afraid that the food would disagree with her. She had reported that to Charlotte, out of the kindest intentions, hoping that she might be induced to change her manner. But Charlotte had only laughed.

It was Charlotte who had introduced her to her husband, calling out as she passed the parlor, "Why, there she is now!" She would have gone on past that day but Charlotte had made her come into the room where he stood in front of the mantel. "Mr. Storrs is Mrs. Trabue's nephew. But we aren't going to tell her he's here till supper time. . . . Her heart's weak." He had given Charlotte a quick look and then they were both laughing. A few minutes later Charlotte left the room, saying she had to see the cook. She herself had been about to get up and leave when he walked over and sat down beside her on the sofa. A tall man, with a coarse, pitted skin and the kindest gray eyes. "Mrs. Mayhew tells me you spend a month with her every summer. She seems to have a very select place . . ." "Yes," she said and in her embarrassment could not keep her voice from shaking. "The boarders are all very nice." And then Charlotte came back and made a fuss over her tan shoes that she was wearing for the first time that day. "I declare, Daphne, you always get the best looking shoes," and he looked at the shoes and then up at her, smiling. "They *are* pretty," he said.

She had wondered then. Why was he troubling to be so amiable? What did he care about her shoes? She had thought that perhaps being lame made him kind, but his limp was very slight. Charlotte said that he had got it falling down a flight of stairs when he was drunk. "It's the last touch," she said, "makes him perfectly irresistible."

At supper that night he sat next to Charlotte and after supper he played bridge with Mrs. Trabue and Mrs. Sloane and Emmie Ronald. Mrs. Trabue had not seemed glad to see him. She was more exacting than usual that evening and the game broke up early.

She and the Rollins girls had gone out on the porch to get cool. When the game was over he came out, stood in the doorway a minute and came right over and sat down beside her. The Rollins girls went inside after a little but he had sat on. He said that the boarding house seemed a strange setting for her. She had got her breath back by that time and asked, "Why?" sharply. "I don't know," he said and laughed. "Perhaps I shouldn't have said that. After all, I don't know you very well."

The next day she went to Charlotte and asked her what she had told Mrs. Trabue's nephew about her. "Why, nothing," Charlotte said, "except that you are General Sumner Wood's granddaughter and play the piano better than anybody in the house. You aren't ashamed of your grandfather, are you?" And then she had burst out laughing. "You can't blame me, Daphne, if he is smitten by your charms."

When she met him in the hall the next day he stopped her, as excited as a little boy, and told her that his aunt had made him a present of two hundred dollars. "It will come in handy," he said. "My little girl's birthday is next week." He was forty-five years old, though he looked younger, and had been divorced from his wife for three years. "The only thing I mind is not seeing the children. I'm allowed to have them with me for two weeks each year. But that doesn't do me much good. A newspaper man can't take care of two little girls . . ."

She had left the boarding house the next week, to go to Chattanooga to stay with her cousins, the Bells. He had followed her there. The job he had been about to take in Nashville hadn't panned out. "They want a copywriter and I'm a layout man, and besides," he said, smiling, "I like the mountains." He had stayed at the Signal Mountain Inn and had come down to see her every night. She knew that the Bells were talking. But she didn't care. Through all his courtship that was not what had worried her. It was the other thing, for *she had known all along*. At every move she had thought, *This is not the way things happen*. But he was there, bringing her flowers, candy, the fitted bag that she had thrown out of the street-car window on her way down the mountain . . .

And the things he had said to her, not his declarations of love, but the little things he had let drop in that low, deprecatory voice. People thought that she had not known anything about his past life when she married him. But he had told her everything, even what his wife had said when they were divorced. "She's right. I'm not any good," and when she had argued that he was, "No. I never will amount to anything. . . . I ought to have met you earlier."

They were married in August. When she asked him where they should go on their wedding journey he said that he did not want to leave the mountains, and wished that she would simply join him at the inn . . . The wedding reception—the house had been heavy with the fragrance of Madonna lilies—was not over till six o'clock. It was dark when they arrived at the inn. As they went up to their suite they passed people coming down to dinner. It was one of the best suites in the hotel: a bedroom, a bath and a large sitting room, facing the Point. While he was tipping the boy she went to the window and looked down into the valley where the lights of the city were coming out, one by one, like the fire-flies now on this lawn. The boy left. She heard Richard's quick, limping step, heard him open and shut a dresser drawer and then the creak of the bed as he sat down. She did not turn around but stood gazing at the lights. But her breath was coming fast and even now when she thought about that time it was as if she held a fiery coal in the palm of each hand. For she had forgotten what she had always known, had believed that she was loved, had almost feared to turn around because now there was no barrier between her and the passion he had professed. And then he had come up and put his arm about her and said that he must trouble her for a loan. He was laughing and his face wore the half-tender, half-mocking expression it wore when he spoke of his wife and children. She had not understood. "They'll take your cheque at the desk," she had told him. He laughed. That was the first time she was really afraid. It was as if somebody else, somebody like Charlotte Mayhew, were in the room, laughing with him. "You don't understand, darling," he said, "I can't write a cheque." His arm tightened about her shoulders. "Aunt Bessie's two hundred just got us up here," he said, and took her chin in his hand and turned her face so that he could look into her eyes. "It was pretty bad of me, I know," he said and laughed a little . . . She remembered making a little mewing sound, looking past him, wondering whether her legs would bear her as far as the bed. She sat down on its edge before she spoke. "I haven't got but twenty-five dollars. I spent it all on clothes. It was foolish of me."

He looked at her intently then. "You could raise some money, though?"

"I don't see how I could. It isn't but seventy-five dollars a month. I get it four times a year."

He said, "Seventy-five dollars a month!" and then he laughed again and said something she had never rightly understood. "Seventy-five dollars a month!" he said. "That little hussy!"

"I could try," she said. "I never asked them to advance me any money before. It's Hollins and Slaughter take care of it for me. I could try."

His eyes had a bright, absent stare. "I wouldn't bother," he said. "We'll get along." He walked to the dresser, picked up one of the brushes he had laid out, ran it over his hair, and re-tied his tie. "I wouldn't bother," he said again in that strange, thoughtful tone and got his hat and moved to the door. He paused there, his hand on the knob. "I've just thought of a friend I can telegraph," he said and went out, closing the door behind him.

She waited until midnight. When he did not come then she went down and told the clerk. Afterwards, when she would lie awake, while the events of that night passed before her like a cinema, she could see that the clerk had repressed a smile as he carefully noted down the name: Richard Storrs. But at the time she did not notice the smile. Two or three times that night she went out of the side door of the hotel and walked about in the grounds. Beyond the smooth lawns, off towards the Point, there was a tangle of wild growth. She had had the delusion that he was wandering there, had kept thinking how a desperate man might plunge off those flat, tilting rocks two thousand feet down into the valley. . . . He must even then have been on the train for Atlanta. . . .

From the next room came the sound of a foot on the floor, a chair pushed back sharply. Catherine burned a lamp till all hours, reading in bed, or perhaps she was writing to her husband. It was strange no letters came from him. At least there had never been any when she brought in the mail. Perhaps they had quarrelled. Catherine was carrying on now with Tom Manigault. Perhaps she had carried on with other men in New York and Jim Chapman had put his foot down. That was probably the reason she had come down here.

The noises had ceased. "How quiet it is now," she thought. "But then it always is quiet here." She went to the window. Maria's cabin was dark. Negroes went to bed with the chickens. The new moon, tangled in the boughs of the sycamore tree, shone, even through the fog. They say it is bad luck to look at the new moon through the boughs of a tree. But I have never had any luck but bad, she thought. . . . I ought not to have thought of those old times.

Seventy-five dollars a month! . . . That little hussy! . . . Play the queen, Daphne! Play the queen and see what happens. No wonder Mrs. Mayhew laughed. . . . But it has taken you twenty years to see the joke.

I did not deserve that. I am a good woman. I have always done my duty.

But nobody has ever loved you. The children used to leave you standing at one side when they played 'Snap the Whip.' Nobody wanted to take your hand.

Agnes loved me.

That was because she had so much life herself. Few people have warm breath to spare. They are afraid you will chill the little breath they have. Mrs. Mayhew couldn't bear to touch the mushroom but you liked it.

Charlotte Mayhew was irresponsible, wicked. She played me, as she would have finessed a queen at cards, married me, for a joke, to a man who wanted only the money he thought I had.

But she was alive. . . . Go to Mrs. Trabue, to Mrs. Sloane and tell your story. They will cry out against her but when the door has closed behind you they will look into each other's eyes and laugh, and Charlotte Mayhew, irresponsible, malicious, wicked Charlotte, in her grave, will be more alive than you have ever been.

XIV

After catherine had helped her grandmother undress and get into bed, she sat down beside the bed and read to her. After a little the old lady told her to stop. Her enunciation was too rapid, she said, reminded her of a guinea's chatter, and besides, the light attracted moths. Catherine blew the lamp out and went out on the porch. Her lips still curved in a smile as she thought of what her grandmother had said about the guinea's chatter. It was characteristic of her. She was growing deaf and when she could not hear plainly she always maintained that the fault lay in the enunciation of the speaker.

The heavy mist that rose from the creek almost every evening floated at the end of the lawn. Tree trunks showed black through it and here and there on its edge fire-flies hovered, wan swimmers, preparing to dive into a cold forest pool. Night fell fast here. Somewhere out in the mist a twig snapped. Any minute now Cousin Daphne would be stumbling in. If she stayed on the porch she would have to talk with her. She went quickly up the stairs to her room.

A letter that had come this morning from Molly Ware was lying on the table. She glanced at it again. Molly thought that she ought to come home and reminded her that she could not disappear without having people talk. She had ended by saying, "Please write to me, Kit."

Catherine sat down before the old secretary to answer the letter but words did not come easily. She rose and went to a dresser drawer where she kept a carton of cigarettes. As she pulled a package out the carton slipped forward, revealing her husband's letter crumpled behind it. It was the only letter she had received from him since she had been here. She had intended to burn it and then at the last minute had thrust it into the drawer. She picked it up and, sitting down on the bed, read it through:

"I want to talk with you. I suppose you would rather I didn't come there. Won't you come here? I will leave the apartment if you want me to, or you can stay with Molly or at a hotel, whatever you prefer. . . ."

She had been out at the box with Aunt Willy when the carrier handed the mail over and she saw on an envelope the familiar, cramped scholar's handwriting. She had put the letter in her pocket without opening it. Even after she was alone in her room she had delayed opening it. She hardly knew what she had expected or hoped to find, but it had not been this. She had achieved a measure of calm during the weeks that had elapsed since she had seen him. But these dry words had been like a blow in the face. The sense of loss, of horror swept over her again. She had answered at once, telling him that she had nothing to say to him and did not want to see him and had driven in to Carthage to get the letter off in the evening mail.

It had seemed to her then that she had no choice, that his letter demanded only one answer. But who can answer a letter truly? When a trivial question is asked one can command one's whole self in reply: we will go to dinner on Thursday night, we will not make a certain journey. But when the question is fatal the self seems to lose its identity or rather it splits into many selves, which engage in a Protean struggle for mastery. We may abhor the grinning devil who dispatched the note that broke up a lifelong friendship or shattered a happy marriage, but no matter. In the moment that he became master he assumed our mask. We may call as loudly as we please but our friend, our lover will not answer, will hear only him.

But if I had to write it over again I would write the same thing, she thought, and got up and walked about the room.

From the blur of shadows over the mantel a face looked impassively down at her, a solemn child's face, framed in dark curls. The little girl of eight or ten sat sidewise in a rope swing, her white, lacy skirts stretched out stiffly. One hand clasped a doll to her breast, with the other she steadied herself in the swing. Cousin Bessie, who had died forty years ago of eating too many ripe peaches and ever since had been staring from that rope swing. I might go to New York now, she thought. I would not commit myself to anything by going. I would simply talk with him. After all, if I am going to get a divorce we will have to see each other again, to arrange things. Besides, as Molly says, it is childish to run away.

She had reached the other end of the room in her rapid strides. Seen from here Cousin Bessie's face was not quite so impassive. The black eyes held pin-points of light, the pouting lower lip seemed instinct with secret knowledge. You can't go—at least until Willy gets back. You promised her that you would stay here. This is the first trip, the first pleasure she has had in years. You can't take it away from her.

Of course I can't, she thought quietly. That's settled. I have to stay here till she gets back.

She paused beside the table. The bottle that held the sleeping tablets she had been taking for several weeks was almost empty. She must remember to get it filled the next time she drove into Carthage. She might ask the druggist if he didn't have a better soporific. This medicine put her to sleep quickly but it was a sleep from which she woke fatigued, and last night she had had a grisly dream.

She had descended, with another woman, into a long, dark tunnel. A man, whose relationship to her was not defined, had walked between them, resting a hand on the shoulder of each. The hand was cold. That was because the man was dead, or had been dead and now, called back from the grave, hovered between life and death. Ahead of them in the vast, shadowy tunnel other people were busy with certain operations. When they had finished those operations, the man would be consigned to another grave, from which, it was hoped, he would rise. But in the meantime he walked beside her and kept his frail, intolerable hand on her shoulder. She was about to shake it off when somebody on ahead called back to her that she must be vigilant, that the man's safety depended on her alone.

She shivered and, drawing a chair up to the window, sat down, leaning her arms on the sill. The mist was creeping up over the lawn. The trunks of the oaks were gray, wreaths trailed from their branches. On the right the mist faded in a watery blur. The new moon, shining dimly, cast a wash of yellow light on the roof of Maria's cabin. Under it the cabin was dark. Uncle Joe and Maria had been asleep for hours, no doubt.

When she had first come here Maria had used to regard her with a peculiar attentiveness. It had annoyed her. She had fancied sometimes that the negro woman could read her thoughts. But now Maria's gaze slid over her as indifferently as it slid over the others, and she answered her as she answered them, only after an interval and absently, as if recalled from a distance. She was continually brooding over her boy in the penitentiary, Willy said. I am afraid her heart is broken, Willy said.

If your heart were broken, if a great fissure came in the center of your being, you might turn your vision inward, might from then on contemplate only what could be seen in those shadowy depths. People in the outer world would become ghostly. You would hear their voices faintly and sometimes forget to answer, as a man, in the midst of an earthquake, standing on the edge of a precipice and gazing down into the earth's heaving bowels, might not hearken to the voice of a companion, calling out to know the extent of the disaster.

A wisp of mist swam across the moon, obscuring its outline, leaving only a smoky blur. I hope it's good weather at the Fair, she thought; I hope Willy has that much luck, for once in her life.

There was a noise on the stairs. Cousin Daphne had come in and was going to her room. Her door closed. There was the sound of quick steps for several minutes and then silence. What does she do when she is alone at night? she thought, and tried to recall the details of Daphne's abortive marriage. Her husband had abandoned her in a hotel on her wedding night. She had heard that when she was a child and it had stuck in her mind. But what could have induced him to marry her in the first place, and, having taken such a step, why had he left her so abruptly? She can't have been very different then from what she is now, she thought, and yet she was young. Even *she* was young once! How could she endure the shock, the humiliation? Perhaps that is why she is so queer. Perhaps that explains her sudden advances, those embarrassing confidences, that odd, jangled laugh. She is always thinking that she is over it, that she is sound again, and then it all comes back and she knows she never will be.

For a moment she was overcome with compassion. It seemed to her that everybody, that she herself, was like Daphne, half-crushed by some early misfortune and having to advance, maimed, through life, cutting in the very struggle to maintain balance a ridiculous figure that somewhere must provoke mirth. And then she felt a revulsion against the woman, against the very house that harbored her. It is this place, she thought. There was always something wrong with it. And in the silence she listened and heard the sound of the creek and it seemed to her that in its babble it announced its purpose, of flowing in a great circle about the farm, of cutting it off from the rest of the world.

The fog was thicker. Here and there a gray bough showed, but most of the lawn was white and over the sycamore only a faint, yellow haze told where the moon had hung. Now that the moon was gone the mist shone whiter. It rolled up to the window. She leaned out. The white mist receded. She felt only cool, moist air on her face. But a few feet away was the other element, into which one might plunge, as into the sea. She felt a desire to be immersed in it. If the stallion were here she would slip him out of his stall and ride out into it. Nobody else would be on the road, she thought. I could let him pick his way. And then she thought that if she telephoned Tom Manigault he would bring two horses and they could go for a ride together.

She went to the closet, put on a pair of slacks and a dark sweater and pulled an old cap down on her hair. She had decided to call the Manigault house. There was a chance that Tom would answer the telephone himself. If he did she would tell him what she wanted. If Mrs. Manigault or one of the servants answered she could leave a message: she was missing Red and would be obliged if they could send a horse over tomorrow afternoon.

She slipped her shoes off and, carrying them in her hand, opened the door and went softly out into the hall. Light still showed from under Daphne's door. She advanced past it on tip-toe, with frequent waits between steps. At the foot of the stairs she turned back through the hall towards the ell. The door between the ell and the hall was open. She closed it and groped her way to the telephone.

When Tom answered she said, "This is Catherine," and was going on to tell him what she wanted but got only as far as "I'd like to go for a ride," when he interrupted her. "Yes," he said quietly and a little absently, as if he already knew what was in her mind. "Yes. I'll be right over."

She had the presence of mind to remind him not to come to the house. "I'll walk up the lane and meet you," she said.

"Yes," he said, still in his preoccupied tone. "I'll be there in a few minutes."

She went out the back door and started up the path to the

lane. A dim shape confronted her. She put her hand out and felt blunt thorns among cool leaves and knew that she had passed the Japanese quince bush that grew beside the path. Her eyes became accustomed to the dark. She made out the gray bulk of the stable looming on her right. Once she had passed it she could find no other landmark and fetched up against the fence twice before she finally opened the gates and stepped out into the soft dust of the lane.

She walked down the lane. Ahead, on the left, a faint, yellow gleam revealed a cluster of dark shapes: the Shannon house and its out-buildings. Mr. Shannon had got into the habit of burning a light all night during his wife's last illness and now that she had been dead ten years still kept the light burning. Mr. Shannon was at the Fair, with Willy. But one of his tenants was probably staying in the house, "to look after things." She went more softly, letting her feet in leather moccasins sink deep into the dust. Here the fog was not quite so thick. Trees were tall on each side of the road. Between them where the lane turned into the highway the mist hung white. A long shaft of light suddenly pierced the white. Boughs sprang out black, light slid from leaf to leaf; a car was turning into the lane. She ran up the bank and slipped behind a tree: a chestnut-oak that had young shoots springing up waist high around it. The sound of the motor grew louder. She knelt behind the thick leaves and watched while the yellow fan swept the dust, then, with an effort, averted her face. What am I doing here, on this dark road, in this thick mist, out of the sound of human voices, out of reach of help? There are drunk men in that car, negro men, coming home from carousing at some pig-stand. They saw my eyeballs flash then. In a minute they will stop the car and come up over the bank and I will hear them asking each other which way I ran. The car was slowing down. With infinite caution, while her eyes still darted among the trunks through which in a second she would take her flight, she turned her head. The yellow light paled, cut by a sharper gleam. A tiny flame flared from a lighter held in the driver's hand. She sprang to her feet, calling, "Tom!"

He got out of the car and took a few uncertain steps. "Where are you?" he asked in a low voice.

She ran down the bank. Catching his arm in both her hands she dug her fingers into the hard flesh while she pressed up against him. "I thought you would be riding," she whispered, "I thought you would be riding!"

He pulled his arm away from her clutching fingers and, setting both hands about her waist, half lifted her into the car. "I was afraid," he said. "This blasted fog. Horse might break a leg."

She did not answer. Sitting quietly beside him, feeling his arm heavy across her cold shoulder blades while the warmth from his body struck through her thin linen garments she let her eyes rove the circle of light that, washing breast high against the tree trunks, pushed the mist back on every side. He leaned over and switched out the lights. He took her chin in his free hand and murmured. She pressed closer to him. "Yes," she said.

XV

It rained in the night and turned cool. In the morning the sky was a deep blue. On the rank green grass the shadows of the maple leaves lay broad, almost black, yet edged, it seemed, so brilliant was the sun, with light. After the midday dinner the women came out and sat on the porch. Catherine sat with them a few minutes, then went to the garden to gather the butter beans for the morrow's dinner.

The path she followed was one that she had cut herself, through an almost tropical tangle of morning glory vines. Uncle Joe had been in bed for two weeks and Rodney was working at the Manigaults'; there was nobody to keep the morning glories down. Maria, for the last few mornings, had refused to go into the garden to gather the vegetables. It was too hot, she said, to gather them in the noonday sun and a person, venturing in there before the dew had dried, risked dew poison, and she showed Catherine her apron, streaked with green juices and dripping with dew.

"I'll gather them," Catherine had told her and every morning since then she had gone into the garden immediately after breakfast.

This morning, however, she had slept late and the butter beans and beets and cabbages had gone ungathered. On a day like this it was really pleasanter to gather them after the dew had dried, she thought, as she moved slowly down the rows, stripping the thick, green pods from the clustering stems. Everywhere the morning glories, climbing on stout okra stems, on other weeds, on the poles set for beans, erected their wreathed tents. It was always like this. Each year Uncle Joe put in a garden large enough for three families, cultivated it for several weeks and then abandoned it to the morning glories. One might have thought them the main crop and the okra, the beets, the onions, showing here and there through the swaying wreaths, the weeds that the gardener had not had time to eliminate. She remembered a garden she had had one year on a small, terraced plot, overlooking the Mediterranean. Not a weed had showed itself through the season. The only menace had been the snails which every morning were found, half dissolved, flattened against the ring of arsenous powder with which every night she outlined the plot.

The basket was half full of beans. She stopped and rooted up half a dozen onions. One of the swollen, green stalks broke. She wiped the juice off on her rough apron, then held her hand up to her face and breathed in the rank odor, and thought: I am going to live in the country the rest of my life.

She left the row and went towards the gate. The house came in view. The women were still sitting on the porch. A bough of the maple tree, hanging low, framed them in a green bower. The shadows that flecked their white skirts were only paler leaves from the same bough. The sun streaks that darted in and out among the shadows edged every leaf with light and spilled in a white sheen on the hard earth below the steps.

She looked away over the lawn. The trees rose out of broad stretches of sunlight. Their topmost boughs had the same sheen, as if the leaves were on the verge of dissolving in the bright air. She glanced back at the house. One of the women had leaned over to retrieve a book. Something else moved in the corner next to the door. But there were only two women sitting on the porch! The third woman was a hundred miles away. She stared into the corner and saw a drift of shadows sway again in the slight breeze that had sprung up.

She passed through the gate and moved slowly across the lawn. Her head a little tilted back, she took the barely perceptible wind on her cheeks, while her eyes roved the upper boughs or fell to the gray clapboards.

She told herself that she had never seen the old house wear so cheerful a look. It was, she supposed, haunted. At least it always had been for her. Often when she was a child, playing in one of the quiet rooms or out on the lawn under the trees she had suddenly been made to feel by some stirring of the air or change of light that she was not alone, and to preserve her privacy she would turn her back or go off to play in another place. The presences had been then only companions whom one could not conveniently address. After she became a woman they had seemed at times to menace or at least to prophesy evil. Four nights ago their voices had driven her out of the house, into the fog, into the arms, she recalled, with a secret smile, of her lover. This morning, as she walked in the bright, scarcely stirring air across the lawn towards the women on the porch the presences spoke in the tones they had used in childhood, but so compellingly that she stopped and with a quick, sidewise turn of the head, looked behind her. There seemed actually to be a stirring of the air all about her, a gentle, continuous murmur that would have resolved itself into speech had there been an ear fine enough to catch it. The illusion was so vivid that she stood motionless, and as she gazed about her it seemed that all the broad stretches of sunlight were crowded with the presences. But they were not here to warn or implore. Their voices were kind, their aspects gay. If they moved under the trees it was to dance.

She told herself that there were no presences, no audible cries, that it was her own mood, coupled with the fine weather, that made everything wear so kind an air, and as she loitered across the grass she reflected on her plight. She wondered why she had always felt that certain actions which she constantly condoned in her friends were not for her. Mona Gilbert had told her that she had only begun to live after she divorced Harry and married Bill. I was married young, too, she thought. Maybe I've only begun to live now and it seemed to her, looking back, that her marriage had been only a long straining to live up to what her husband demanded of her. She recalled evenings when she had sat silent for hours while two men, or half a dozen, conducted a conversation of which only an occasional phrase was intelligible to her. It had always been like that. When she had the opportunity of making new friends—and she rarely had that opportunity nowadays—her first thought was not whether she liked the new acquaintances or whether they liked her but whether they would be acceptable to her husband. Even now, after fifteen years of married life, she could not tell what would bore him. Her old girlhood friend, Molly Ware—a rattlepate if there ever was one—was one of the people he liked to have around. He listened to everything

Molly said and often had long discussions with her. But there was Sylvia Brice, who had a good mind and would have liked to talk with Jim. You could hardly get him to stay in the same room with her, and if he did stay, if he did present the appearance of listening, it was with what Molly called his "glassy look" and his replies were so inept, so disconnected that you felt sure that he was following his own train of thought all the time.

If she were to marry Tom Manigault—and she was at the moment determined to marry him—her life would be very different from what it had been with Jim. They would live in the country. There would be the succession of country pleasures which so absorbed and delighted her. It is the life I was made for, she thought, the life I have always missed. She thought of Tom. Since that night of the fog she had met him alone twice, each time in the woods. That first day when she had come along the path he was standing under the oak tree but he did not come to meet her. He stood and gazed, as intently as he had gazed when he first saw her, that day she rode over to Big Pond. He did not speak or move until she was opposite him, then he said:

"I stayed awake all last night, wondering if it was true."

She dropped down on the bank, laughing. He came and stood over her. "I turned on the light," he said. "I saw this," and he pulled out of his pocket the old cap she had asked him to stow there on the night of the fog. "I saw this," he said, "and I knew it was so."

She laughed again, taking the cap out of his slightly reluctant hand. "You're like a puppy," she said. "You give a puppy your old shirt to sleep on and he'll stop howling."

Later, for without any other word he had set his hand on

her throat and borne her down, still half laughing, among the dead leaves, later, when they sat side by side, silent, subdued, watching the bright spears of sunlight strike from leaf to leaf, down on to the brown floor of this secret place, she had reminded him that she was ten years older than he was. "How will you like having an old, haggard wife?" she asked.

"I don't care if you're as old as Methuselah," he said. "We are going to get married. . . . We *are* going to get married, aren't we?" he repeated and abruptly turned and looked at her. The strange yellow flecks in his eyes had given them an almost metallic gleam. But there was something in his look besides resolution, an apprehension that amounted almost to terror. It had hurt her to see him look that way and she had put her hand out quickly to cover his. "Yes, we're going to get married," she said. "I'll go to Reno. As soon as Willy gets back."

She was at the porch. She put her basket down and dropped down on the top step.

"There are your butter beans," she said.

Mrs. Lewis turned her massive head on which the white, curling hair still grew vigorously. Her blue eyes were growing dim but their gaze, suddenly withdrawn from the distance, was intense. Of late she sat often and stared in a kind of fierce bemusement. What was she staring at?

She was leaning over to pinch a bean. "They're filled out," she said. "Sometimes Maria'll pick beans before they're filled out."

The dog had rushed into the hall to lap water from his bowl and now rushed out again. The old lady drew her white linen skirts aside as he pushed past her and sniffed a little.

"That dog has been rolling in something," she said.

Catherine leaned over and sniffed, too, then straightening up, drew the dog to her. "Oh, no," she said, "He smells sweet." She held him against her, erect on his hind legs, and intoned in a high, squeaky voice:

"Sometimes I roll in manure,
Just to increase my glamour . . ."

Cousin Daphne's black eyes were bright. "Is that a poem?" she enquired.

"He often writes poems. The first poem he ever wrote was an ode to his mother. It says:

If you loved me as I love you,
You'd treat me a whole lot better'n you do . . .

"Does he write many poems?" Cousin Daphne asked.

"He hasn't written many since he's been down here. He isn't really a Nature poet. But when he's in town . . ." She fell silent, watching the dog pad down the steps to settle himself in the soft dust between two maple roots. The old ladies had been shocked by Heros' poem but they liked it. . . . Jim always liked Heros' poems. They couldn't be too silly for him. He even liked him to make puns. When he got home in the evening he always asked what Heros had said during the day, and if he hadn't "talked" she had to make something up. . . . But he hadn't talked now for a long time. This was the first time he'd spoken since he'd been down here. She put the thought of her husband out of her mind. "It's because I'm happy," she thought. "He always talks when I'm happy . . ."

The floor boards creaked as the old lady rose. Catherine got up quickly and fetched her cane from where it stood in the corner. The old lady stood a moment, leaning on it heavily and looking down at her cousin. Daphne lay back in her rocker, her legs hooked over a rung of a straight chair drawn up in front of her. The old lady seemed to be eyeing the thick ankles encased in black cotton stockings. Daphne, as if aware that the scrutiny was unflattering, withdrew her feet from the other chair and sat up straighter.

The old woman laughed. "You sit like all the Passavants," she said.

She went slowly across the porch and down the hall. Catherine followed her into her room. Old Catherine sat down in the broad-bottomed rocker which stood always beside her bed, and with a little groan, bent and began untying her shoelaces. Catherine dropped on the floor in front of her. "Let me do it, Mammy," she said.

She quickly untied the lace and withdrew the foot from the soft, black shoe. She had taken both shoes off and was about to slip her grandmother's feet into bedroom slippers, when the old woman suddenly elevated her legs and held them stiffly out before her. The feet, encased in white cotton stockings, looked not so much like feet as mittens, stuffed, Catherine thought, with cotton batting, ready to hang on some child's Christmas tree.

"Do your feet hurt, Mammy?" she asked.

Mrs. Lewis shook her head, and then with uncharacteristic frankness: "I wish they did. Sometimes I feel like I'd give anything to have some feeling in them, even if it was pain."

"It's not your feet, though. It's your ankles, isn't it?"

The old woman stood up. "It's all dead wood. Somebody ought to come along with an axe and lop it off."

She held her arms out like a child while Catherine unbuttoned her shirtwaist and slipped her skirt over her head. Catherine laid the garments on a chair while her grandmother, in her old-fashioned chemise, went behind the screen to wash her face and hands.

She took a long time. Catherine sat down and waited, hands folded in her lap. From behind the screen came sighs and splashings. Catherine thought of the morning bath in which the old lady sat firmly planted on a stool while a basin of soapy water and a wash cloth were extended to her at intervals by the attendant who must lurk on the other side of the screen. "It wouldn't be so hard if she weren't so damned modest," she thought. "How does Willy stand it, day after day?" Her hands moved nervously. She folded them again. "I must be like that," she thought, "I must do everything as well as I can."

She lifted her head and as she gazed about the dim room it seemed to her that if she were not as attentive to her grandmother, as patient as Willy herself, her new happiness might be menaced. Already the exhilaration that had come to her in the last few minutes was fading. It was the effect of this room, where the blinds were nearly always drawn, in the morning because the old lady's toilet was still in progress, in the afternoon to exclude the heat of the sun. Outdoors it was still light. As soon as her grandmother was settled in bed she would announce to Daphne that she was going for a walk. She had promised to meet Tom in the woods. "It's absurd that I can't have him at the house," she thought. Still, it doesn't matter. I like it better in the woods. . . . It's going to be all right. Everything is going to be all right. I'm going to be happy. There isn't any reason why I can't be happy. . . .

The old woman walked heavily across the floor and climbed into bed. Catherine approached and drew the sheet up as high as her breast, then bent and kissed her lightly on the brow. "Have a good sleep," she said.

The old woman, lying on her feather pillow, did not acknowledge the caress. She stared at the wall. Catherine was halfway across the room before she spoke. "When I was eighteen I could span my ankle with my finger and thumb."

Catherine laughed. "Heavens, Mammy! I bet you were vain."

The old woman's dark gaze still fixed the faded wall paper. She did not answer. After a moment Catherine closed the door and stepped out into the hall.

The old woman slipped down between the sheets, turning her body sidewise until the feathers hollowed out under her left hip, then brought one of her hands up and laid it beside her cheek on the cool pillow.

In the hall she could hear Catherine's voice in the high, maudlin tones she affected when she was "speaking" for the dachshund.

She gets it from the Lewises, she thought. It is a habit that runs in families. I'm glad John never had it. I could not have borne it. No, I could not have borne that.

Catherine was going upstairs, accompanied by the dog. *Mother . . . Mother . . . Mother.* Long ago she had heard her father comment on the fact that childless couples often addressed each other as "Mother" and "Father," as if recognizing the existence of an imaginary child. But Catherine made a dog call her mother. She wondered if the girl were barren. Probably not. People sometimes took a notion not to have children nowadays and if they took the notion they could keep from having them, it seemed. Agnes had probably got on to those new-fangled ways, too, after she went to New York to live; she had never had but the one child. It might be that women didn't have to do anything nowadays, that they didn't have the children in them. Catherine was thirty-five years old but she acted more like a girl than a woman. Perhaps when a woman didn't mature her seed didn't ripen either. In time there might not be any more children. The race might vanish from the earth.

She turned her head fretfully on her pillow. "I must be a little feverish," she thought. "Such notions!"

Catherine had come downstairs and was going out with the dog, who was speaking French now. *"Irons plus au bois, Mother. Irons plus au bois. . . . Beaucoup des lauriers."* Catherine was telling him not to be so free, but to address her as *Madame ma mére.*

The old woman shivered and held her hand out before her, the gnarled fingers spread at first and then, bending inward, seeming to clutch at something.

I have lived a long time and have seen those I loved die and have known other troubles and other sorrows and still there are two sounds I cannot endure: the cawing of crows in a high wind over an open field, human words put into the mouth of a beast.

In the hall the light, ironical voice called out once, then died away.

She is in the woods now, she thought, on that path that goes to the spring. But there is some one there before her, the woman who is always on that path. Sometimes she is lying face down and her fingers rustle the leaves. Sometimes she stoops over, gathering into her apron pieces of dead wood, broken twigs, dry leaves. . . . *That is because I came that way the first time and the last time too. . . .* She is moving aside. I never saw her move aside before. It is for Catherine! Her granddaughter was stepping too lightly past the stooping woman. She knew anger, thinking of the poor hearth on which the woman would fan the broken twigs, the dry leaves to a blaze. And then it seemed to her that her granddaughter had gone on and the stooping woman straightened up and the path was hard and brown under her feet and above the open field that could be glimpsed through a break in the woods, a crow flew towards the sun, cawing.

"I hate to hear crows caw," Sue Robinson said. "It always sounds so sad."

"It sounds mad and glad, you fool, but what good is it when your voice whines on and your sallow face turns to us at every bend in the path?"

Ned was thinking the same thing. He looked at her and then he looked at me and smiled and all the light came to a point in his eyes.

I thought, he will be gone tomorrow and I hurried and caught up with her. "If you will read to Aunt Jessie this afternoon I will do it three times running," I said. She stared and the corners of her mouth drew down. But my hand was on her arm. I pushed her. "If you don't I'll never let you wear my filigree necklace again. Never!"

We stood there and watched her go off down the path. And I was frightened at what I had done. "She will tell everybody," I said. "She will tell everybody we are off here in the woods by ourselves and would not let her stay with us."

He did not say anything till she was out of sight and then he whispered *"Kit!"* and I forgot how I was not going to let him kiss me—the girls all said he was such a flirt—and we sat down on a log by the path and I let him kiss me until I feared I was all tousled and drew back. He put his head down in my lap then. I could feel his lips hot through my thin, summer skirts and I brought my hand down over his black hair and I said, "Ned, please," and he straightened up. "I won't be here tomorrow," he said.

They went to Camp Dick Taylor first. There were thirty rode off that day. I kept his old setter, Molly, for him. She thought she was going too. "You stay here with Miss Kit. . . ." She was wagging her tail long after they were out of sight. . . . I was the first girl in the neighborhood to get a letter. He said he wore the scarf I knitted for him all the time. The boys were teasing him because it got dirty. He washed it out sometimes in a branch or pond, he said, but he did not take it off except for that. I got more letters than any girl in the neighborhood. . . . Until '64. The letters did not get through then. . . .

In the summer Joe Torrant came home. He had lost his

leg and the other foot was gangrenous. They propped him in the doorway where the breeze came. He kept singing:

A life on the Vicksburg bluff,
A home in the trenches deep,
Where we've dodged Yank shells enough,
And our old pea bread won't keep . . .

There was a sutler named Logan, he said, and he wished he had cut his heart out before he left there, while he could still get around. *Pea bread . . . Pea bread . . . Our old pea bread won't keep. . . .* You laid a piece down and it would crawl away, Joe said. . . . It was not so hard going without the letters. I knew he was at Vicksburg and we were all waiting. I thought all I had to do was wait. And then Vicksburg fell. "We may not win the war," Uncle Cleve said, "I never thought to say it but we may not win the war."

Mama went over and caught him by the shoulder and shook him. "You old fool! You sit there and wear out pants while better men are dying. We *can't* lose!" She raised her head and glared around and the same trees were in the yard and the sun was on the lowest step the way it always was that time of day and I thought, "We *can't* lose!" There was a wet spot on the floor where spit had fallen out of Uncle Cleve's mouth. Mama was sitting down and her head was in her hands and she was crying. "I've struck your father's brother. An old man. I don't know what got into me. I think I'm going crazy."

Ned was at home then. Joe Torrant told us. He had his goats trained to a wagon by that time and he stopped at the gate and asked if we heard any news from Cousin John's folks.

"Not for some time," Mama said. "How are they?"

"They can't do anything with Ned," he said.

I stood where I was in the hall but Mama went down the steps. She kept her voice low but I heard. "Is he hurt much?"

He looked at his stump stretched out before him. "He feels it more than I do," he said.

I did not want to hear it from him. I went upstairs and lay down. The plaster in the ceiling was cracked the way the ground gets after it has stood under water too long. If I could only get through the afternoon! He will be here by supper time, I thought. Something has happened and he could not get away, but he will be here by supper time.

After a while Mama came and stood in the doorway. Ned and John Lewis had both been home several days, she said. She didn't know why we hadn't heard. John was all right but a bullet had gone through Ned's neck. His face was paralyzed. He couldn't talk plain.

I was looking at the ceiling. If I could get my hands into the largest crack and tear hard enough it would all come down. . . . *Certainly I can come down to supper. I am perfectly well.*

Joe Torrant was still there. And John had walked over from Oak Quarter. Joe would not talk about Ned before him. It was while he was in the house he told us how Ned went off to the creek every day. *"But he won't let anybody go with him."*

I looked up and I saw the evening star over the cedar and I knew it would be light for a long time yet. I got up and went in the house. John was coming through the hall but he did not say anything to me. I went out the back way and slipped around through the tall weeds. The old dog went with me.

I stood there on the path a long time and then I heard something move off down towards the creek. There is a bluff there above a deep hole. They sit on the bank and drop their lines over. He was standing under the sycamore tree. The pole was lying on the ground and he had the line in his hand, trying to put the bait on. But his hands shook and the line jumped back and forth as if it had a fish on the end of it.

The path was grown up in sumac. I had to keep pushing the branches aside. He heard me and turned around. His face was there, with the leaves all around it, shaking, the lips stretched flat over the teeth, the eyes glazed and then the glaze went away; the eyes grew fierce. The line that he still held came up between us in a great loop. He flung it from him and took hold of the trunk of the tree with both hands, his face still turned towards me and I knew he was holding on to the tree to keep from shaking, and I was through the sumac, I would have been with him in another second, when he cried out.

The first time it was a sound such as children make when they say, "No, I can't talk. I am this beast or that beast," and they go "Gobble . . . Gobble . . ." "Gobble . . . gobble" they will go. It was only that, a kind, pleading sound. But when I did not answer and ran on he cried out again. I was not afraid of him. Before God I was not afraid of his look or of that deeper, harsher sound he was making. It was his upraised hand. I knew that when he struck me it would still be shaking and I could not bear it. I put my own hand up over my face. I turned back.

The dog crashed past me. I came out on the path by the spring. Sun still trembled on that big rock, on the ferns that pushed up around it. I fell face down, I tasted earth, I clawed at the dead leaves. There was no feeling in me, except in those fingers pushing the dry leaves aside, digging deeper into the damp earth, moles that, having endured the light of day for a second, must seek their holes underground. I think I went to sleep. It was dark and I was cold and stiff and my fingers had not moved for a long time when I stood up.

There was a noise down by the ravine. The dog was on the path, barking. "He is going home now," I thought. "She is going with him, the way a dog goes with you, circling two miles to your one. I will not have the care of her any longer," I thought and I fell to laughing, alone in the woods.

I laughed a long while, not caring who heard me and then I got up and went home. I went another way from the way I had come, through a field that was grown up in milkweed and goldenrod. I lost the path and did not try to find it again but went through the weeds that were breast high and rank with dew. The moon was out. I stood still in the middle of the field and the white light was over everything and I saw the house black under the trees and I did not want to go back to it.

They were sitting on the porch. I came up the walk and sat down on the bench in the shadow of the vines. My skirts were all dark with dew. Nobody said anything about it. I did not go upstairs to change; I sat and talked with them until Mama told Joe Torrant which room he was to sleep in and said she was going in to bed.

When they had left I got up and sat down on the top step of the porch. "The sun dries things. Maybe the moon will dry them, too," I said and spread my skirts out around me.

John came and sat beside me. He put out his hand and touched my wet skirt. "You came through the field," he said. "You ought not to have come through those weeds."

"It was the shortest way home," I said.

He said: "The next time you want to go for a walk, let me know. It is dangerous for you to go about by yourself. Promise you will let me know."

"I promise," I said.

The next time he came he brought me Ned's letter. I saw the backing: *Miss Catherine Fearson. Kindness of Bro. Jno.* and I hated the letters that could talk so surely and steadily and I would not have read it but John was looking at me; so I read it through and when I had finished it I said, "I do not think that this letter needs to be preserved," and I tore it into four pieces and threw them into the fireplace. We were sitting in the parlor. They lay there among the dead ashes all the time we were talking but before John left he pushed them over against a live coal with the toe of his boot and we stood there and watched them flare up and turn gray.

"That is what you wanted me to do, isn't it?" he said, and I said, "That is what I wanted you to do."

I did not go to Oak Quarter for a long time after John and I were married. John said I had better not. "It would be painful for you and painful for Brother Ned."

He would not come where there was company. He lived

in the office. His sister, Martha, used to go down and sit with him. He would write to her, on the tablet they kept hanging by the door, but he would not speak. But he would talk to the old setter, Molly. Martha would sit there all afternoon, with her work, and he would be talking to the old bitch.

Jack was three years old that day we went past the office on our way to the mulberry grove. Ned was standing in the doorway as we came down the path. He had on some old nankeen trousers and above his soiled waistcoast he wore my scarf. Jack saw the old dog and broke from me. I stooped and had hold of his hand when I heard over my head the sounds and, still stooping, looked past the man's legs into the room. It smelled of mildew and leather. There were two beds—the dog slept on one, they said—and the table was piled high with men's and horses' gear. They could clean it up, I thought. They need not let him live like that. I straightened up and the sounds had stopped. Martha was smiling.

"Molly says leave the boy here while you gather the mulberries."

I looked at Ned. His eyes were bright and he smiled, looking down at the child, squatting in the dust, his arms around the old dog, and sounds broke from him again.

"She'll take good care of him," Martha said and we walked on.

The mulberry grove was on a rise in the middle of a field, a wind blowing through it and the ground clean except where the yellow leaves and the black mulberries were fallen.

"He makes the dog talk for him!" Belle Rogers said.

"He did it when he was a boy," Martha said. "We used to have an old gray cat. Sat at the table. At least they put a stool for him. . . . Gimme *'at* piece. . . . Gimme *'at* piece! . . . and if Mama said anything the boys would loose those horrible cat-calls and say Tobe's feelings were hurt. I don't see how Mama stood it. . . ."

"It sounds so queer when he can't talk plain."

"I can understand him," Martha said. "I understand every word he speaks."

"But does he ever say anything to you? They say he always puts everything into the dog's mouth."

"Will you be visiting in the neighborhood much longer?" Martha asked and her voice was cold as she straightened up, resting the full basket of mulberries against her hip. My basket was full, too. I walked away from them. I could see the field through the black trunks. It was sown in wheat. Yellow on top, but where the wind ran were caves, red like earth. And the men were in them, moving around or just standing, resting their guns on top of the embankment.

The Yanks would stop for breakfast and an hour at noon and again at sunset, but except for that the noise went on all the time. Regular, like axes clearing new-ground. The Minié balls were nailing on the shingles, Jim Waldrop said. Ain't they got the house up yet? he said every morning. Ain't they got the house up yet? Oh, listen to the Minié balls, Oh, listen to the Minié balls. And the Minié balls are singing in the air. . . . *They would not ask for a flag of truce, so their dead went unburied. There were as many wounded as dead. . . . I let that boy from the mountains go, once, with water. He wanted to go back. There was a man kept stirring, trying to brain himself with his musket. He made me go to the parapet and look. A Minié ball took his cap off.* Get back, you fool! . . . *That was on the thirtieth. Their parallels were within seventy-five feet of our works; the only thing was to counter-mine. We had seven shovels. But Joe Torrant took three men and went to the hill and cut hickory staves and sharpened them. I kept fifty men on the night shift. In the morning when I went into the trench I stepped in their vomit. It was hotter when the sun got up. That was the hottest day of all, but enough breeze to carry the smell. My sweat burned my eyes. When I went past the bucket I would throw a dipper of water in them. It took a minute to clear away; I heard the courier before I saw him. . . . General Pemberton. . . .* Oh, listen to the Parrott shells, Oh, listen to the Parrott shells and the Parrott shells are whistling in the air. . . . *I stood up. The river shone like brass and the sky was high and bright. The man who had been trying to brain himself had stopped moving. It is better for it not to rain, I said, the sun will finish them up quicker, I said and I followed the courier out of the trenches.*

Major Ross was in there with the general but he went away. The general's hair is turning gray. He talks as if there were more than us two in the tent. . . . Safety in numbers, though I do not doubt that Captain Wise has gotten through, in which case General Johnston will move immediately. . . . *You cannot smell them from here. A woodpecker has just lit on the tent-pole. There are still birds! . . . If I leave them it is to get help. Joe said five rounds. . . .* I have heard that you are an excellent swimmer. . . . *Yes, General, yes. Certainly I can get through. . . .* You have a good memory?

. . . *Yes, General.* . . . Then I will not commit the message to writing. . . .

There was not much shelling that night: the moon was not shining. I was at the mouth of the creek when the masked battery opened up. . . . It took the transport a long time to burn. I heard the bushes crackling around me and knew that they had changed the range and I slipped under the branches of a water-oak and floated out into the river. There were no gun-boats, only bluffs steeper and blacker than ordinary and over my head the sky was rosy like it is before dawn and the water had little cool currents in it. . . . If I left them it was to get help. Joe said five rounds. . . . *And then the sky was gray again and I turned over and struck out free and it got darker and the cool currents bubbled up all around me and each stroke I took was stronger than the last. . . . But you could not swim on through the cool water. You had to get up on the bank.*

The horse was cropping grass on the edge of the woods. I had just enough rope for a bridle. I kept to the woods as long as I could. It was a wagon road I came out on; you would not have thought it was guarded. The one who saw me first had a handkerchief tied over his mouth to keep the dust out. He fired too quick. It was the man behind him got me. When I opened my eyes they were not there. The officer had an apple clenched in his teeth when he bent to feel inside my shirt. It was the trickle of green juice set my chest on fire and then the flames started up everywhere. After a while, when they were burning steadily, he was still there. . . . Yes, he is badly hurt, but I think he can talk. . . . You will save yourself a great deal of trouble. . . . *His lips were moist under a brown moustache. Through the burning I could feel his hand moving inside my shirt. It is not there. It is inside and now it will never get out, for it is burning up. . . . I held my lips flat against my teeth and sent my eyes deep into his and spoke the words that made him start and fall back while the shaken bushes rained gray dust all around us, the words that nobody can understand, that nobody can bear to hear, the only words I will ever speak from now until the time I die.*

She moved restlessly on her pillow. It is a long time ago, she thought, and I did not love him any better than John, only in a different way. But there is not time to think of that now. There is only the one thing to think of, ever. *They have been here all the time.* I cannot think why I never saw them before. They have been with me all along! Sometimes their shadows fell ahead of mine, and even then I did not notice. . . . Until that day when she was standing there. I cannot remember what went before but it was early morning and I was coming up the walk. There was dew on the bricks. And she was standing there in a white morning robe on the back gallery. She had her head bent a little to one side and her hair was all fallen down. She did not look at me but gazed straight ahead from under the dark hair and I thought how she would stand, doing some small thing, gazing, while a whole morning went by, and I called to her sharply. "Have you picked the raspberries?" But she did not turn, she did not answer, and I went towards her. "They ought to have been picked while the dew was on them," I said and reminded her of how many people there were in the house and how much there was to do, and then somebody somewhere else said something and I tried to answer but all the time I knew I must look at her. But I could not see her face plainly for the hair. . . . And then she turned, she was looking at me. . . . I cannot tell how her look was or what she would have said, but I would have known in a minute. I would have known, but a person was beside me on the walk, touching my shoulders, my arms. I said "Alice?" and then I knew by the touch of the hands that it was not Alice and I was afraid. But Willy was there and she put her arms around me and read to me and all the rest of that day it was just like it is every day. It has only happened once and it may never happen again. . . . It will come some other way. But it does not matter what way it comes. It is there, waiting. It has been there all the time; I cannot think why I never knew it before. And you cannot tell whether it will just be being with them or something else. . . . I did not mind her gaze or the dark way her hair fell. It was the moment when she was not there and the house and the trees were not there, either, and I did not know who I was or who the person was who kept calling.

When John died I did not think about it. I would turn over in the night and cry for the emptiness. And then I would be under the earth with him. But he could not look at me for the earth stopping his eyes. It was the earth heavy on his eyes that I could not bear. And the worms. . . . I never thought of the wandering. . . .

For that will be the worst thing. And it will be in a place like no place you have ever seen and people will come and go and never answer and you will not know where you are.

. . . *I know that my Redeemer liveth. . . . Hail, Mary, full of Grace, Blessed art thou among women. . . .* Aunt Rosa was a devout Catholic and died with her rosary in her hand, but if they hadn't sent her to the Sacred Heart Convent she would have been a Presbyterian, like all the Fearsons.

How lovely the crab apple is in the spring! In old times, when they cut down the trees, they always left it standing; "the lady tree" they called it. That little island has been there in the creek ever since I was a child. But then the wild geraniums grew down into the water. The water has washed them all away. It is deeper than it used to be. If it were not so deep I would cross over and rest in the shade of the lady tree. But the water is deeper than I remember and off in the woods there are people, laughing. . . . *He has been there all the time, smiling at me through the leaves!* He is coming out from under the boughs, his hand is stretched out. . . . But there is some one here beside me. . . . I cannot stay with you, Ned. John is calling. But you may take my hand. . . . Is this Pigeon Creek? It is deeper than I remembered. If we slipped and went down, the people in the woods would not know but would keep on laughing. . . . The mist is rising. I cannot see your face for the mist but I know the touch of your hand. . . . There is so much mist. If we do not hurry we will lose him. . . . No, he is there, still smiling. And the leaves are green all around him. . . . Will you come up and rest with us, on the bank, among all the green leaves?

XVI

DAPHNE CALLED to Catherine as she was leaving the house, but she did not answer and went on towards the woods. Tom was there before her. He had ridden in from the upper road and had thrown the reins over his mare's neck and turned her loose to crop the long, fragile grass that grew in the ravine.

Catherine exclaimed when she saw the mare. "She'll go home," she said. "They'll think you've been killed and scour the country."

He laughed. "No, she won't," he said and whistled. The mare stopped grazing and took a step towards them, her mild eye fixing his. When he did not whistle again her head sank and she moved forward, slowly cropping grass. "I trained her myself," he said. "Comes like a dog."

"Did you break her?"

"She didn't need any breaking. I just slipped a bridle on her one day and threw my leg over her and we rode down to the mail box. All that High Cloud stock is gentle. . . . You want a colt out of her?"

"I'll have to think about it," Catherine said. "I believe I like Daisy better."

"That's because you don't know one horse from another," he said.

"I couldn't know very much about them," Catherine said meekly. "I live on East Sixty-fifth Street. We only have room for a bicycle under the stairs."

"God, I'd hate to live on East Sixty-fifth Street!" he said.

She did not answer, running her hands through the long grass that came up all around them. "You will often find *Agaricus silvicola* in this woods grass," she said. "There is another mushroom that much resembles it, distinguished by the floccose brown scales which ornament the pileus. . . ."

"That Cousin Daphne is too much for me. You say she eats 'em?"

"We had a tasty little mess of *Trompète du Mort* last night."

"Trumpet of Death?" he said, "I don't want you eating that kind of thing. You hear? I don't want you eating any more of them."

He had thrown himself on his back on the grass and now he stretched his arms up and clasping them behind Catherine's neck, made her bend and kiss him; then would have drawn her down beside him but she put her hand on his shoulder and gently pushed him back. "I'm out of breath," she said. "I had a terrible time, getting away from Cousin Daphne."

"Anybody would think you were three years old, the way they treat you," he said.

"I'm thirty-five," she said, evenly, looking over his head up the ravine, where the gray beech trunks rose out of the tall grass. A few of the beeches were already showing yellow leaves. The lively color did not seem a presage of Autumn but only an added touch of opulence in the summer green. It is a Giorgione landscape, she thought. The trees, the luminous shadows, the horse cropping grass and the warrior, recumbent, with his mistress beside him. Only

I ought to be naked and more voluptuous looking. What is the matter with me, she thought, that I view this scene as if I were a beholder and not an actor? Has Jim infected me so that I must see everything through his eyes?

He was sitting up. "Have you got your breath back?" he asked, smiling and looking into her eyes. She felt his hand on her waist, moving her body slowly towards his and she put her arm up about his neck, drawing his head down till their lips met in a long kiss and then, shuddering with the suddenly communicated desire, she stretched herself on the earth and looking up saw over the crisp, sun-burned hairs of his head the sky blue against green leaves before she closed her eyes against the too ardent gaze and was alone on the sandy beach and the cruel, green, foaming waves plunging hard and the beach quiet until, roused by the strokes, it curves upwards and they cling together in the cool, climbing, ocean-beaming air before the jealous sea draws the wave back and the beach, relinquished, lies quiet under the moon.

He had turned a little away from her and lay on his back, his head resting on his upstretched arm. She pressed against him, thrusting her head into the hollow under his arm. The arm descended and gently pressed her closer. She shut her eyes. There was no sound except the minute singing of some insect hanging on a blade of grass. The grasses that their bodies had crushed gave off a sweet smell. We ought to be lying in a bed, she thought, but it is only tall grass. Leave them lying on the bracken, leave them lying where they fell, better bier ye cannot fashion, none becomes them half so well. . . . It is not a bier. It is our marriage bed. And I do not mind its being in the woods, where any one can come and surprise us. It is you who are amazed, my dear. We are surprised if only everything is all right, if only everything is all right. She sat up and put her hands to her hair and drew a silver box out of her pocket and looked into the tiny mirror and powdered her face.

The reins had slipped down over the mare's head and the snake-like thongs whipped in a circle about her hooves as she grazed. He went over and adjusted the reins on her neck, then came back and sat down beside Catherine. He put his hand over hers.

"I wish we could go away from here," he said.

"It's only three more days," she said. "Willy will be back on the twelfth. I'll leave the next day."

"You haven't written Chapman?" he asked abruptly.

"I haven't told him that I'm going to Reno."

"Why haven't you?"

She made an impatient gesture. "I haven't been communicating with him. I've written him only once since I've been down here. After all, he isn't particularly concerned about the date." Or is he? she asked herself. I wonder if he wants to marry that girl?

"I think you ought to tell him," he said.

"I'll write tonight," she said.

He was silent. "What's going on at Big Pond?" she asked. "Do you realize I haven't been to see you in over a week? . . . Do you suppose they'll get suspicious of us, Tom?"

"I don't want you to come there," he said. "At least till there's been a change."

"What do you mean?"

"I'm not going to have that Miller there much longer," he said abruptly.

"What's he done?"

"I went in the kitchen last night, to get a drink of water, and there was Roy, sitting up on a stool and drying glasses for Rodney and the two of 'em laughing and talking."

"I don't see anything wrong with that. Tom, you know yourself you like to talk to niggers. And Roy is awfully curious, about how people live and everything like that."

"I like to talk to niggers, too, but I don't like to sleep with 'em." He turned on her a flushed, distorted face. "A man that'd debauch nigger boys!"

"I don't believe it," she said deliberately. "I don't think he'd do a thing like that, down here, when you've all been so kind to him."

He laughed harshly. "You can't trust 'em, I tell you. They're not going to let anything stand between them and their pleasures. Mother knows all about it. She just won't admit it."

"I should think it would be hard for her to admit it," she said. "After all she comes of a different generation from ours. I doubt if she knows what a fairy is."

"She lived in France a long time," he said.

"Yes, but she frequented high society. That Frenchman who stayed with you all at Sands Point. . . . It was perfectly obvious, but he was the son of that princess and nobody said anything about it. . . ."

He was silent. She leaned forward and touched his arm gently. "Don't be upset," she said. "He'll be going away soon. It'll be much better if you don't have a row."

He did not say anything for several minutes, then he said in a low, constrained tone, "If she didn't have him she'd get somebody else."

"You mean you think she keeps Miller around just to annoy you?"

He did not answer. "Tom?" she urged gently.

He raised his head and looked into her eyes, but his gaze, even while it met hers, was withdrawn, self-absorbed. A slight, secretive smile trembled on his lips. "You don't know," he said.

"I know that she's a very vain, spoiled woman. And I don't think she's ever taken the proper kind of interest in you."

"She takes an interest in me, all right. That's why she's down here. She can't stand to be away from me."

She could not think of anything to say. He was kicking at the earth with his booted foot. The mask-like smile widened. If he lifted his eyes to hers they would still have that cloudy, withdrawn look. She tightened her clasp on his arm. "Tom, what do you mean?"

He looked up. His eyes brightened. He laughed, without mirth. "It was when I was about eight years old. Out at Sands Point. She and the old man had been quarrelling and I came in after he'd left the house. She was lying there, looking out to sea, and she turned around when I came in and she looked at me. She looked at me and I knew what she was thinking."

"That she hated you?"

He nodded. "I don't know what they'd been quarrelling about. Maybe she had more principles then than she has now. Maybe she felt she couldn't divorce the old man because of me. I don't know what it was, but I stood there and let her look at me until finally she turned away and her handkerchief whipped up in her hands and she told me to call her maid. But I didn't do it. I just walked out of the room."

"They say women often hate their children," she said softly.

"Well, when anybody starts hating me I hate them back. I started hating her when I was eight years old."

The mare stopped grazing and came over and stood in front of them. Her wide look said that it was time to go home. Catherine made an unconscious gesture and the animal, repelled by the light motion of the hand, walked away. The land is not enough for him, Catherine thought, or his beasts or his friends or the women he will love. There was something went before. And it is no use to tell him that his mother is not one to awaken either love or hate. The enchantment fell too early and her features are the most glamorous that he will ever see and he will always be harking back to that brightest moment of danger.

She felt tired. "I'm sorry," she murmured and gave his unresponding arm a quick pressure, and let her hands fall in her lap. It must be four o'clock, she thought, the light has changed. She looked up the ravine. The light had changed, but not towards evening. It was brighter than she had ever seen it before in the woods. Every leaf glittered with light. And the hush was oppressive, as if some power somewhere had abruptly stilled all sounds. I have made a mistake, she thought, I have taken the wrong road and it is too late to turn back. Am I lost? She trembled. It seemed to her that she was alone in the woods and the glittering light had a voice, a voice that would have spoken but for the command laid upon it, and then suddenly the stilled woods gave tongue and the air all around was quivering with the wild, high-pitched, despairing cry that brought her to her feet and sent her racing towards the house.

XVII

"It was dreadful for Catherine," Mrs. Manigault said.

"But how fortunate that Tom happened to come by," Miller remarked.

She gave him a quick glance. "He wasn't there at the time?"

"No. Cousin Daphne found her, although you would think it was Rodney, to hear him tell it." Miller raised his voice to a falsetto. " 'Old Miss laying there in a pool of blood, and didn't know nobody nor nothing, and never will know nothing, the doctor say.' "

"I wish you didn't find it necessary to roll your eyes when you imitate Rodney," Mrs. Manigault said. "Did the doctor really say that?"

"It was a brain hemorrhage, accompanied by complete loss of memory. She may get her memory back, of course. She may live on for years."

They had dined on the terrace and after the table was cleared had sat on in the dusk. Mrs. Manigault looked back now into the lighted house. Through the window she could see Tom sitting at his desk. He had gone in after dinner, saying he had some accounts to figure and had been there at the desk ever since. A few minutes ago she had seen him close the book in which he kept his accounts. He had not risen from the desk but sat smoking and staring at the wall.

"Tom?" she called and saw him slowly turn his head. When he did not rise she called him again. "Tom, do you suppose I ought to go over to the Lewises? Is there anything I could do?"

He got up and came and stood in the doorway. "Catherine said she was going to nurse her herself. Said Cousin Willy wouldn't want them to get a nurse. And she's not going to telegraph Willy."

"No," she said thoughtfully. "That poor Willy! It was dreary enough before. And now she will have to play nursemaid for the rest of her life."

"The old lady may not live long," Miller said cheerfully.

"She'll live for *years,*" she said with conviction. "There's nothing improves the health like loss of memory."

"I suppose so," he said. "Takes the nut off. And all your physical energies flower. I expect the old lady'll be a handful."

She glanced at Tom. She had thought that he was going to come out and take a seat on the terrace but he continued to stand silently in the doorway for a few minutes, then with a muttered "Good night" turned back into the hall.

"You're going to bed early," she called.

He did not answer. She heard his feet ascending the stairs and then the sounds were muted on the rugs of the upper hall and she heard his door close softly. Miller sighed and stretched luxuriously on his long chair. "This has been a beautiful day," he said. "I looked out over the east field this morning and you could see the air just tumbling."

"Tumbling?" she said.

"For *joie de vivre*. I wanted to catch it in a cup. . . . Elsie, have you ever eaten peacock?"

"No," she said.

"Rodney tells me that in the old days that was the regulation Sunday dinner around here. I suppose it's more like a guinea than anything else. Dry, and a bit gamey. He says his mother bakes them to perfection. It seems that the Robinsons continued to have peafowl on their table long after the white people gave them up. When Rodney was a little boy there were a lot of old peafowls straggling around, didn't hardly know who they belonged to."

"I suppose that's what became of the flock that used to belong here. I had to go up in Kentucky to get mine."

"I suppose so. *Baked*," he repeated reflectively. "It would be sure to be dry. But there is no proper roasting these days."

"You are not going to eat any of my peacocks," Mrs. Manigault said.

"No . . oo. Well, there are the guineas. I think I will teach Clara a new sauce. Didn't you think the guinea hen was a bit dry last night?"

"I didn't touch it," she said. "I'm dieting."

"I don't know why. Your figure is perfect."

They were silent, watching the poplars throw their heavy spears of shadow across the rectangular sward.

"Do you think Tom is really interested in that girl?" she asked abruptly.

"He is evidently much interested in her. But whether they contemplate marriage . . ." He gave the little sibilant laugh she had come to dislike. "I am no expert in affairs of the heart."

"You're a man," she said bluntly. "You ought to know something about how another man feels. She's too old for him but I think he's in love with her. But I don't know whether she'll marry him. After all, Jim Chapman is a very distinguished man."

"But Tom has so much money."

She sat up straighter. "I happen to know that he has a note coming due at the First National Bank this month," she said. "Why they keep on lending him the money is more than I know. . . ."

"Feudal feeling. They like to see a Manigault farming Big Pond." He sat up too and leaned towards her. "Elsie . . ." he said.

Hearing his voice sink on her name she turned her head quickly, staring. *"What?"* she asked.

He swung his legs over and sat on the edge of the long chair. His hands, extended before him, the fingers a little curved in, seemed in the dusky light to suspend between them some fragile, invisible object. "Elsie," he said, "why don't you go away from here?"

"And turn the place over to Tom?" she asked in the same low tone.

"Yes," he said eagerly.

Her eyes were on the fragile, invisible shape that his hands enclosed. "What would I do then?" she asked crisply and saw his hands sink slowly to his sides and in her fancy heard from the flagstones a brittle tinkle. "What would I do?" she repeated.

"You could go to New York. Take a pent-house apartment. Do something. . . . There are so many things you could do. . . ."

"What?" she reiterated.

"God, woman, how do I know? . . . You could start an

art gallery. You could buy some pictures and open a small gallery."

"And you would tell me what pictures to buy?"

"Naturally. You could start with two or three. Unknowns. I know where to lay my hands on one now. There's a Dutchman living on Hudson Street, a spiritual descendant of Seurat, and, mind you, Seurat died when he was thirty-two. His work was just begun. This man understands Seurat. He understands Poussin, too. He's got a new theory of Divisionism. God, Elsie, doesn't that excite you?"

"No," she said. "I don't want a gallery. They would paint the pictures and you would tell me which ones to buy. I wouldn't have any part in it—except to furnish the money."

"The sinews of war. Isn't that enough?"

"No. I've got to be doing something myself."

He was stretching himself back on his couch. "Women are all wrong nowadays," he said. "You ought to be more like Isabella d'Este . . . or Vittoria Colonna. . . . I wonder that you've never gone in for politics."

She pursued her own train of thought. "It must be my father in me. I'm not happy unless I'm doing something."

"I can't understand it," he said. "If you had to sweat for your daily bread it would be different, or if you were driven to create beautiful things, like this house which I called into being out of nothing," he said with a proud wave of his white hand. "There are only two reasonable forms of effort: those that contribute to survival and those that contribute to art, which is a deeper form of survival."

"I was decorated by the French government for my work in the hospital," she said.

"Naturally. You were handsome, energetic, rich, well-connected."

"They took the horses out and set me on the gun carriage and we paraded down the avenue. . . ."

"There is nothing like war," he said kindly. "We are all artists then."

She was silent, hardly hearing his words, thinking of the crisp, black, springing hairs that, except for the shaven areas surrounding the swollen, empurpled wounds, covered the whole body of Paul Bourdue. His was the first bed in the row, but the men would not have him bathed first. Even old Raoul, whose lips murmured continually over the fingers tented over his chest—he said five hundred and eighty Hail Marys and six hundred Pater Nosters while he was in the hospital—even old Raoul when she came to his bedside would smile and ask if the basin contained fresh water. "I do not wish to catch his fleas, Madame." And Paul's black, Midi eyes would shine and he would maintain that he bred only lice, and one day when he caught one in the matted fur of his chest he held it up for all the ward to see. "A prize winning hog. Fattened in the Forest of Bourdue." He always wanted her to bathe him. Sister Victorine could never learn to roll the sheet out right. . . .

". . . I do not wholly agree with Tolstoi. It is not merely the relief from moral responsibility. It is the escape from self. Every time man makes a sacrifice he rids himself of a part of his burden. . . . But you will have the opportunity again. It looks as if we were getting into the war. By February, I should say. . . .

No, she said silently. No. There will never again be a

time like that and I do not want it. I am not as young as I was, and I have a son. No, I do not want war . . .

". . . And I think you pacifists, you women, are to some degree responsible for the present crisis. We have been feeding the youth of the land lies, for a generation now. And when was youth ever anything but cannon fodder?" he asked sadly.

She stood up. "Have you a detective story in your room?"

"There's an Agatha Christie on the table by the bed. And there are three or four in the bookcase. I recommend *The Spider in the Cup* which you will find on the third shelf from the bottom, third book in the row."

"I think I'll take the Agatha Christie," she said, "if you're through with it."

"Oh, quite," he said courteously. He stood up, too. As she passed him he bent and lightly kissed her cheek.

She was gone. He stared into the doorway where her body had showed for a moment against the light. In the last few weeks he had fancied that she was beginning to look her age. She was as slim as ever, but she did not carry herself as buoyantly as she once had. She had stepped up into the doorway, heavily, like a person suddenly lifting a weight. Gravity, which all her movements had once seemed to defy, was tugging at her. The flat thighs, the long, shapely legs, the small, round breasts had begun their long declension towards the earth in which they must eventually rest.

He sighed and lit a cigarette and lay back on his couch. He thought: Women . . . they live longer than men but they do not wear as well. I am glad I kissed her then. I am the only one to kiss her these days. And there will not be anybody after me. . . . He half closed his eyes and tried to summon her image as she had been when he first knew her. She is the woman I have come nearest to loving in my life, he thought, and I must try to remember her as she was, must try never to forget. It is little enough that we can do for each other in this life . . .

He opened his eyes, disturbed by a sudden shifting of the light which poured from the broad façade of the house out on to the greensward. It fell just short of the trees that reared themselves at the end of the enclosure. The tall poplars rose out of a band of black shadow. But the shadow was turning gray. Some one walked behind the poplar boughs, carrying a light.

He has finished his work, he thought, *and now he is taking the garbage out to the incinerator*.

The incinerator stood, like a marble faun or a fountain, at the end of an alley of pleached apple trees. In the daytime the grass under the trees would be molten with the interlaced, quivering shadows, but now the way was dark except for the beams cast by the flashlight. The man on the couch followed with his eyes the gleams flickering from bough to bough. The dark boy would have arrived at the end of the path. There were no more gleams of light. He had laid his flashlight down in the grass. He was bending to turn the handle which would release the metal door. His feet were planted wide apart. He bent lithely from the waist. Under his white houseboy's coat his brown body was wet with sweat.

He half rose from his long chair, then sank back. It always comes like this, he thought, after a time when I have had peace and am lying in the sun and the forces that I thought had departed forever are rising in me. I could have worked on my book this morning, but I did not feel quite enough power and I thought I would wait another day . . . That has happened before, he thought wearily. It was Robert's fault that I have not finished my book before this. I never worked better than I did at Nassau. I had the whole of the second section in my head ready to write down that morning. My manuscript was spread out. I was working by the tall window. Julio was arranging flowers on a table when Robert came in at the far door and shouted. It was a long room, but he need not have shouted so that Julio, going out by the other door, heard him . . . I never saw Julio again . . . When I think of that time now it is not Julio that I see but Robert, backed up against the wall, shouting. Two folds came at the corners of his mouth and kept drawing them down, and I thought how ugly he looked. I never thought until that moment that he might want me for himself. But who could have wanted him when Julio was there. How ugly he looked, backed up against the wall, shouting!

Deliberately he contemplated the image which for a time had been graven so sharply on his memory that he could never think of the moment without making some unconscious gesture of revulsion. But the lines that fixed the grin of rage on Robert Askew's face seemed stereotyped and without power to move and the grin itself faded, leaving the features of his former partner smooth, youthful, impassive.

He brought the leg which had been hanging over the chair up to lie beside its fellow. He folded his hands on his chest. I must go carefully, he thought. Elsie is my oldest friend.

Tom is almost a son to me. I must go carefully. His thoughts formed themselves into words, but the words seemed to be spoken without conviction by some one standing at a great distance. He repeated them, striving to make them take on life. "I must go carefully. I must not let anything like that happen here."

The light was flickering on the poplar leaves again. "He is taking the pail back," he thought. "He has set it on the steps, and has turned down into the basement. He is taking off his white coat and his dark trousers. He is standing there naked under the shower. . . ."

He got up and walked to the end of the terrace. He hesitated, glancing back over his shoulder at the quiet house, then took the path that led between low box bushes to the latticed basement door.

Upstairs Mrs. Manigault reached over and switched out the lamp beside her bed. As she lay quiet, her eyes closed, she became aware of a small throbbing in her arm. The cut she had got that morning when she backed suddenly against a strand of barbed wire was hurting again. She got up and went into the bathroom and took a bottle down from the cabinet. She poured a drop of iodine into the wound, then held her arm out, rigid, while the dark liquid spread and filled the tiny furrow. I won't bandage it, she thought, cuts always get infected if you put iodine on them and then bandage them.

She stood at the east window a moment in her long satin night dress, waiting for the iodine to dry so that it should not soil the sheet; then, getting back into bed, pulled the blanket and a sheet up over her. The book she had been reading lay open, face down on the table. She hesitated, then picked it up and read a few pages. But the conversation going on between Poirot and Hastings did not hold her attention. She laid the book down and, stretching her body its full length, lay back upon the pillow, her head resting on her upraised, folded arms.

She had been standing in the corner of the pasture, holding the bay mare on a slack bridle rein. She had not seen the white paper blow across the grass. Her eyes were on a group of colts at the other end of the pasture when the rein suddenly went taut and out of the corner of her eyes she had seen the white-stockinged foot flash forward and in the same instant had seen the lower lip gaping to reveal greenish, parted teeth. She had backed up against the fence, so suddenly that the mare's head had been jerked around. They had confronted each other. The mare was actually rearing when Joe Marble ran up. Standing close beside the mare he had grasped the bridle with his right hand while his left arm slid over the withers and down the back in a long, quieting caress. She had had the feeling that the little man, if he cared to, could have picked the mare up and set her down in another quarter of the field.

But he had been agitated. His eyes shone. He spoke harshly. "Don't ever let her see you're afraid of her, ma'am!"

The barb had gone so deep into her arm that she had had to have help from one of the negro boys before she could extricate herself.

"I wasn't afraid," she said when she went over to stand beside him. "I wasn't in the least *afraid*. She startled me."

He stroked the mare's quivering nose and spoke to her

in his quiet voice before he let one of the boys lead her away. "No'm," he said, "but it 'us a bad idea, letting her back you up against the fence."

"I think I'll have that barbed wire taken down," she told him.

He had agreed that barbed wire was responsible for a great many accidents and they had made plans to replace the old fence by one electrified wire. She had almost forgotten the incident in her interest in the new fence, but now, lying alone in the quiet room, it came to her as events of the day just lived through often come back in the night. In the dark she felt her face flushing and she set her teeth for a moment upon her lower lip. She had cut a ridiculous figure, hung up there on the barbed wire, with the mare rearing and about to lash out with her fore feet while all the stupid negroes stood staring.

Her head groom was an under-sized, spare, dark man who moved unobtrusively, speaking seldom and always in a low voice. Until today she had never seen him agitated. But the face he had turned on her today was quick with anger—and accusation. She had had for a second the odd feeling that the anger had been there all along, that he would always have spoken to her in that harsh, authoritative voice but for the patience that like rubber, coating fiery electric current, overlaid his every movement and his least utterance.

"But he is always burning inside," she thought. "That is why he handles the boys so well. He hardly ever gives them an order and always in that quiet tone, but they jump for him faster than they ever do for Tom."

He had had only one employer before he came to her: a Colonel Edward Andrews in the next county. Tom had asked him once how long he had worked for Colonel Andrews.

"I started when I was four. Exercising the horses. They had a big nigger was so lazy he couldn't hardly draw his breath and he'd put a bitting rig on a horse and put me up and turn us loose on the track and every time we'd come past him he'd wake up and say, 'That's right, son, keep on riding him!' Least," he added, "that's what my daddy said. He come down there one day and found me taking old Saint Blaze around and he went up to the house and told the colonel, 'See here, Colonel. You going to work my boy you ought to pay him.' And the colonel give me a dime every Saturday. I remember that because I bought 'Long Jim' with it," he added reflectively, "and seems like I remember Old Blaze. But you know, I can't remember that nigger at all. And it may not any of it be so. It was just the way they told it."

She had gone to his house once at night, to consult with his wife about something. He had remained in his chair, reading *The Breeder's Gazette* until the women had finished their talk, when he got up and came slowly towards her. "I don't think we better . . ." he began and she knew he was referring to the breeding problem they had discussed earlier in the day. He had not thought of anything else during the interim. He did not ever think about anything but horses. He lived in a world in which horses were the only objects worthy of contemplation. When he relapsed into those periods of abstraction the negro grooms stepped softly about him, whispering to each other, "Don't bother Mister Joe, he thinking," but he was not pondering his financial condition or his domestic affairs, he was not wishing that he had taken a job somewhere else or that she would pay him more money; he was working out some problem of breeding or handling or showing. When she had approached him with her offer of a house and garden and a wage of a hundred and fifty dollars a month he had looked abstracted and she had thought that the amount was not enough and was about to raise it when he said with a quick, dismissing wave of the hand, "You better talk to my wife about that. . . . You say you got a High Cloud brood mare?"

That is what it is to be an artist, she thought. Roy is always talking about how superior the artist is to the common man, how everybody else is born for the whip. I wish I had been born like that. I wish I cared about some one thing more than anything else. But I have not ever really cared about any of the things I have done. . . . The furniture factory. At first it seemed such a good idea and I liked the smell of the wood and I would go and stand beside old Mr. Naylor and watch the lathe turning and I worked hard in the office and was patient with those women and would call Mr. Naylor in and we would consult about how you could change the desk to make it fit into a space three by four and then one day that fat woman came in and we were looking at the desk and I saw her lip wrying and I thought, What is the use of making beautiful reproductions for them when they would rather have something from Grand Rapids? And I did not tell Mr. Naylor for another month, but I knew I would have to sell all the equipment and find him another job. I could not go on with it. . . . Helen Turner says that I have never found my *métier,* that I ought to work with people. If I had started with one of the big department stores I could have a big executive position now, she says. . . . For a moment she saw herself seated behind a gleaming desk in the spacious, sun-lit office. A tall, dark young man stood in front of the desk, head slightly bent until, leaning back in her chair, she began speaking, when he raised his head and looked keenly into her eyes, nodding his head at intervals. A young secretary, white frilled, in a dark dress, noiselessly manipulated a typewriter. But what would be the good, she thought? It is only to make money and I have enough of that. . . . No, I had rather be here. It is the nearest I have ever come, the nearest. And you need not say that it does not make any difference, that you ought to be glad to have food and clothes and a roof over your head. You could do without them sooner. For it is the only thing that matters. . . . That is where all the years have gone. There would be the beginning and the excitement and I would think *This is it. This is it,* I would think, and I will not change now but will go on with it, and then the dead feeling would come and I would have to get out of it. You would think you could make yourself go on, but you can't. And it is not the dead feeling but the time after when you know you never can. And there is nothing to do. Sometimes I have had to leave a place and no reason to give. . . . But there is nowhere to go. . . . This is the nearest I have been. The dead feeling does not come and there are days when I forget who I am and I do not mind exercising the horses when the boys do not show up or getting up at night with the colts. And there is so much to do yet. . . . But Roy says I ought not to stay, that Edward would not want me here. He never showed much affection for the place when he was alive, she thought with anger. He did not seem to care what became of it. It is not right to let a place run down the way this place was run down. I was needed here. That is why I came. And it was all right, at first, when I was living at the hotel and driving out early each morning and the men had started working and each day you could see a little change from the day before. It was all right till Tom came. . . .

He is your only child.

But he is young. There are so many other things he can do. I will give him the money.

This conflict is unnatural. He is all of flesh and blood you have.

I have never had anything for myself. And now it is too late. It would be different if there had ever been anything for me. I could give it up to him then. It is not much, not what I looked forward to, but it is all I will ever get now. I cannot give it up to him. . . .

She sat up in bed. The sounds whose existence she had for some minutes been unconsciously denying had grown louder. There was a single, loud exclamation and then a door closed belowstairs with a loud bang. She slipped into a dressing gown and went on bare feet out into the hall. She leaned over the banister and called "Tom!"

There was the sound of feet sliding on a rug. His face, contorted with anger, showed at the bottom of the stair-well. "Aye God!" he shouted, "you want to come down here?"

Her body stiffened. "If I'm wanted," she said, and still in her bare feet, slowly descended the stairs.

He was not in the hall. She pushed a door open and stood in the little room which he used as his office. Tom was standing beside his desk. Roy Miller stood with his back against the opposite wall.

He saw her. A smile jerked at the corners of his mouth. He made a deprecatory gesture. He addressed her, but his black, shining eyes were fixed on Tom's face. "I'm leaving," he said, "with sound and fury."

"No, you won't!" Tom said. "She's got to know all about it."

"I was surprised with your houseboy in the basement a few minutes ago," Miller said quietly. "Tom, quite rightly, wants me to leave the place. At once," he repeated, and took an erratic step towards the door.

Tom had turned to stare at her. "In the basement," he said. "You want 'em to have cots and showers and everything else in God's name and he's been going there every afternoon, with Rodney."

Miller again half lifted his hand. "There is no need to distress your mother with the details," he said, and lurching oddly, walked past them out of the room.

Tom breathed hoarsely. One of his eyes was blood-shot. "That's what you've had here in the house," he said. "I knew all along. I could tell by the way he kept eyeing him. But you wouldn't listen. No, he was your old friend. . . ."

Through the open door she could see Roy's feet slowly ascending the stairs. Her hand that had been fumbling with the ribbon of her gown, went out stiffly towards Tom. "*Please*. He's going."

He walked the length of the room, then returning, brushed past her and went through the door. Outside on the terrace she could hear him striding up and down.

She went up the stairs. Miller's door was shut. She hesitated, then knocked. There was no answer. She knocked again. A voice said, "Come in."

A travelling bag stood open upon a little stand. He sat in a chair he had drawn up beside it. A suit of clothes, on its hanger, lay across his lap. Neckties swirled in a bright cluster over the chair arm.

He turned his head, made as if to rise, then abruptly leaned over and, taking out his handkerchief, dabbed at his neck, his forehead. He gave a little laugh. "These encounters are very fatiguing," he said.

She continued to stand before him. "Don't take anything except what you need tonight," she said. "I'll have your things sent. . . . Have you any money?"

He shook his head.

She went into her room and wrote two cheques and brought them to him. "You can cash the little one at the hotel," she said.

He raised his face, still gleaming with sweat. He threw the crushed, damp handkerchief from him and stood up and took a step towards her, then, although there was no revulsion in her face, in her eyes sadly fixed on his, he stopped short. He held out his hand for the cheques. "Good-bye, Elsie," he said.

She went back to her room and lay down on her bed.

She lay there quietly until she heard the taxi come up the drive. But after the front door had closed and the house was quiet she got up and walked about the room and finally paused in front of her father's portrait that, in a heavy silver frame, sat always on her desk. He sat on a marble bench, in spring sunshine, the trunk of an oak dark behind him, wearing a suit of rough gray tweed. A bird dog lounged up against him, one paw on his knee. The man sat, a little turned away from the dog, from the spectator, his thin hands, whose mottles of brown she could remember, holding a silver knife with which he was cutting off the end of a cigar. She remembered the day that Sutton, a photographer famous for his studies of men, had come out to take this picture of Papa, at her request. He had posed him among his papers, among his books, beside a statue, under a Bronzino portrait of a duke, "everywhere," the old man had said with his dry laugh, "except at the piano," and finally had taken him out to this bench in the garden. The dog had come rushing up. Eph Cottar, relaxing, about to light a cigar, had thrust a foot out to ward him off. But the photographer had called out for him to let the dog stay and had taken the picture in that instant. He had done well. The handsome cloth, veiling the shrunken limbs, the white marble of the bench, the dog whose pose might signify affection, the silver hair, all had combined to present to the world the picture of an old man serene in the sunshine, his head bent over a trivial task. If he had turned his head you would have seen in a face, so narrow that it might have been drawn with two strokes of a child's crayon, deep-set eyes, still glowing coldly at seventy, a pinched nose, a mouth as thin and cruel as a carp's. As a small child she had been a little afraid of her father's gaze, of his sudden, too wide smile, but familiarity had dulled the first keen impressions. When she came to adolescence and found that she could deceive him, could even impose her will upon him, she had developed an amused affection for him, an affection that had increased after her marriage. He was, she told herself now, the only person who had ever really loved her and she paused before the silent figure and, extending her hands, whispered, "What shall I do?"

He did not answer. Elsie Manigault was his only legitimate issue. The other child, a son, was a mulatto. His birth in a Memphis brothel, his life shuffled out along the river front, his death from a bullet that a sheriff had meant for another man, all fated, set in train by the moment when a group of young men hesitated at the corner of two snowy, ill-lit streets until the whisper, "Dark meat's cheaper," sent them up the nearest rickety wooden steps. Eph Cotter did not know of this son's existence and would not have cared to know about it. That had been a careless, unguarded hour, one of the few he had ever known in his life. Until he was past forty he had had no time for either marrying or whoring, being already dominated by the lust which first took whole possession of him when he was nineteen years old, on the same morning that the Butztown Rifles marched past his father's store on their way to the railway cars that would take them to Manassas Junction. His father, a mild ox of a man with perpetually troubled blue eyes, had stared at the marching youths and then back at his son. "Ain't you going, Eph?" he had asked. Eph, stepping back out of the sunlight and feeling harsh, laden cloth rubbing against his bare chest—the canvas bag his mother had sewed for him at the beginning of the summer, held six silver dollars; three hundred more were hidden under his mattress—Eph had not troubled to answer, only turned on the man his cold, abstract gaze. In imagination he had already quitted his father's roof and was sitting among piles of ill-smelling hides in Charlie Schrader's warehouse, engaged in the interview that would make him manager and part owner of the tannery. That same tannery that from now on would be so mysteriously and persistently mismanaged that Charlie Schrader, frustrated by the strange disorder of the accounts to which he finally got access by candle-light and despairing at the last of a long train of reverses, would end his life with a bullet.

"He was the only one and now he is gone," she whispered. "What shall I do?" she whispered again, raising her eyes to the portrait.

He did not turn. He did not answer. She was the child of his middle years. He had not married until he was forty-five, but even when she was born he had been too wise for such questions. Not with the wisdom of the imagination that comes sometimes to men in their youth, to an Heraclitus, a Buonarotti, a Leonardo, but with the wisdom of an old, tired organism; of the old possum that lies all night in its hole rather than drag its crippled leg over the snow to the henhouse, of the hawk that, discerning with fading eyes the flash of wings through the leaves, yet clings to its perch on the dead pine, preferring the faint pain of hunger which it has had for a long time and will have until it drops shrivelled from the bough, to the agony of the moment when, having struck and missed, it must beat upward through empty blue.

XVIII

When willy left the hotel at seven o'clock the square was already full of people moving towards the park. A light rain had fallen in the afternoon. Puddles of water stood in the concrete; the grasses that thrust up beside the sidewalks were beaded with moisture. Willy hesitated on the steps, then went back upstairs and got the sofa pillow that she had brought from Swan Quarter and thrust it under her arm. She would need it in the grandstand: the seats were sure to be damp. Mr. Shannon had wanted her to sit in a box, with a Mr. and Mrs. Proctor from Lewisburg, but she had told him that she preferred the grandstand.

The sidewalk ended. She crossed a muddy road and set foot on the lighted common. The crowd, which had surged so compactly along the walk, was scattering. People stopped to talk with friends or hurried purposefully towards the ring. On the littered grass two boys were wrestling. She watched her own shadow waver ahead of her and fall across the moving bodies and thought that it looked gaunt and lonely.

She came to the tall elms that guarded the entrance. The boys at the gate knew her now and when she made to open her purse for her pass they stayed her with a genial wave of the hand. She slipped through the turnstile and was in the narrow, sawdust-covered lane between the tiers of seats. In the bright space ahead figures moved slowly, came to a stop. The two-year-old fillies were lining up before the judges' stand. At seven-thirty the two-year-old walking stallion class would be called.

Without a glance at the fillies, she hurried along at the foot of the stands. Box three, where Mr. Shannon had wanted her to sit, was full of people. The small woman in the red hat was probably Mrs. Proctor. The woman, standing with her back turned, talking to her, looked more citified, in a dark gray suit and small purple hat. She was turning around. From under the brim of the purple hat blue eyes swept the crowd. She was turning back, smiling, to say something to the other woman. Willy dodged behind a stout man in a Palm Beach suit and walked on, keeping his body as a shield between her and the occupants of the box. The man was pausing at the command of the woman who was with him. They seemed about to take some vacant seats in the first row. Willy hurried on. At any moment Mrs. Manigault might see her and call to her. The stout man caught up with her. He and his wife had decided against the first row seats. They stopped and began climbing towards some seats in the top row. She climbed with them. "If she sees me now, all I have to do is wave," she thought. "I couldn't get down to her over all those people."

The stout man's wife was putting a layer of newspapers down on the damp boards. She saw Willy's sofa pillow and smiled. "You were smarter than I was," she said. A woman next to her was jogging her arm. She turned and, recognizing a friend, began an animated conversation.

Willy stared out over the rows of heads into the arena, empty now of horses. Merry Girl from the Cherryvale Stock Farm had won in the walking filly class. Two figures had left the judges' stand and were strolling over the turf. There would be an intermission before the next class came in. She hoped it would not be long and for the moment regretted her anonymity. It would be nice to have somebody to talk to while you were waiting. Not all the time, just while you were waiting. Yesterday Red had won the Two-Year-Old Futurity and had also won in the Two-Year-Old Stallion class. Mr. Shannon had refused to become excited. Today would tell the tale, he said. Today he was showing in the All-Age. Some people might think they were fools to enter a two-year-old in the All-Age, but he was going to do it. . . . "There ain't anything here that can touch him. But I don't like that Joe Polk. . . ." Joe Polk was from Lewisburg, and therefore, according to Mr. Shannon, to be feared. "These folks around here used to thinking they got all the horse flesh. They're not going to take to a Pennyry'al horse—if they can help it."

A man had been talking, in a great voice, about bonds. He stopped. The megaphone began a sharper pleading:

"Will the judges come to the arena? Will the judges come to the arena? We want to give you a good show."

Two figures broke from the group in front of the boxes and started across the turf. The man in the light-colored raincoat had been pointed out to her earlier in the day as Joe Polk. He was a lawyer in Lewisburg, with a good practice, and had a farm four miles out of town. He had bred last year's winning stallion, Yellow Boy. Mr. Shannon didn't think so much of Yellow Boy. He came of western stock. Didn't have as much of the old Black Allen in him.

The two men, at the foot of the judges' stand, had stopped

to say something to each other. One brought his clenched fist down lightly on his palm. The other threw his head back. If you had not been so far away you could have heard him laugh. The figures disappeared up the little stairway. The megaphone ceased bawling. The stallions had passed through the east gate. They were coming around the curve now, slowly. A black in the lead, a gray behind him. You could not see Red yet. The stout man leaned forward.

"Twenty-two," he said, "that's Bill Rives' horse."

"I like the gray," his wife said. "*Ooooh,* look at his tail!"

"Too showy," the man said, "got stringy shoulders."

"It's just like a waterfall," the wife said. "John, how do they get their tails to stand up like that?"

"Cut the muscles. . . . Twenty-two. That's Bill's horse."

The black horse walked slower. His neck arched, at the same time swaying ever so little from side to side. Each time his left forefront sought the ground it was with a proud curve. "Ain't he got a nice nod?" the man said. "Nice action all over."

The black had passed. The gray's head was small, but he was well ribbed. His tail, set higher than those of the other horses, sprayed his dappled rump with silver. . . . A dark red muzzle was blotting out the silver plumes. Willy looked once at the thrusting shoulders, at the brim of Mr. Shannon's broad, black hat, then looked away, up the line to where the black had come to a stop in front of the judges.

"You think that black's the best horse?" she asked in a low voice.

The stout man threw away the stump of his cigar and bent towards her. "Yes, ma'am," he said when she had repeated the question. "That's the best stallion here. He won at Fayetteville last week."

The woman reached over and extended a sack half full of peanuts. "I hope the gray wins," she said. "He's got such a beautiful tail."

Reverse and canter, please!

Reverse and canter, please!

The megaphone had bawled twice before the horses turned and started back. The gray was in the lead now and next to him the chestnut, then Red. . . . *And if he wins, Catherine said, it'll be on his walk. His canter's choppy. . . . Who wants to canter, Mr. Shannon asked, when they got this fast walk? I'd a had Miss Catherine show this horse if she wasn't a lady. You put a lady up and half the time the judges are looking at the lady, they ain't looking at the horse. . . . Go on, Mr. Shannon. It's because I don't ride well enough. . . . You got a nice, light hand. . . . But no legs, Mr. Shannon, no legs. . . . Ain't no lady, hardly, got a grip in her legs. I wish I weighed fifty pounds less. But I'll ride him as light as I can. . . .*

Will the riders please put their horses to the walk?

Will the riders please put their horses to the walk?

I have to look now, I have to look. If I don't he may not win. . . . That boy knows how to handle the black. . . . *I'll ride as light as I can.* . . . He does look smaller, up. Maybe it's the broad-brimmed hat, or the gray trousers. He's holding him in a little. That's for the nod. . . . *Red, he don't nod as much as some.* . . . But his shoulders shine so red in the light! . . . From the hoof up, he said. They judge from the hoof up. . . . The Allen walk is pretty. Each time you think the hindfoot'll hit the ground before the forefoot. If he had a longer reach he'd caulk himself. His reach *is* longer than the black's, and he walks faster, and he's got better shoulders. . . . She tried to catch the glance of the bright, dark eye that as he passed her strained a little backwards in its socket. . . . You never walked sweeter, Red! They can give the prize to whoever they want. . . . You never walked sweeter. . . .

The megaphone had been bawling for a long time. Or had it just started? The stout man brought his clenched fist down on the planks. "I'm a monkey," he said, "I'm tee-totally a monkey."

His wife was laughing. "Long as the gray didn't get it I'm glad it was the chestnut."

"It was the chestnut, wasn't it?" Willy whispered. "He did say the chestnut?"

The man looked at her curiously. "Yes, ma'am. . . . Them judges ought to stick their heads under a pump." He stood up. "I'm going down and see what Bill has to say. I bet he's rearing." He looked over to where the stallions were cantering, one by one, through the gate. "He does move nice, that chestnut," he said.

Willy took the program that she had torn in two and arranged the pieces neatly on her lap. "I thought he seemed a little nervous," she said.

When they had gone she sat still for a moment, then climbed down and made her way through the crowds to the owners' quarters. Red had already been rubbed down and blanketed and stood with his head thrust over the gate of his stall, his ears pricked forward, watching the moving, wooden horses of a brightly lighted merry-go-round. Speaking softly, she went up and stood beside him. He jerked his head to one side and nibbled at her sleeve. She gave him the apple she had bought for him that afternoon at the A. and P. and when, in his swift crunching, he let a half fall to the ground, she picked it up and held it flat on her palm for the eager lips to snuff up.

"No doctoring the horses, ma'am!" a voice said.

Mr. Shannon was standing in the door of the tack room. He stepped out, and slipping his hand under her elbow, assisted her over the sill. Two men who had been sitting on a long, tarpaulin-covered box in the corner, stood up.

"This here's the lady that owns Red Allen," Mr. Shannon said.

The two men took their hats off and laid them on the box. "You sure got a horse, Ma'am," one of them said. He picked his hat up and moved towards the door. "I understand Bill Rives ain't feeling so good," he said and emitted a short laugh. "I have to laugh when I think about these Lewisburg folks," he explained. "I'm from Marshall county, but I like to see 'em get taken down. They get to thinking they're the only folks can breed walking horses. I like to see a horse come in from outside every now and then." He paused and held a thick forefinger up, "But not too often, lady," he said, "you don't want to do it too often."

"We'll win the Grand Championship next year," Mr. Shannon said.

The two men laughed and disappeared through the doorway.

"Grand Champion Walking Stallion of the World," Mr. Shannon said softly. "That'll be old Red. Next year."

"Grand champion," Willy said faintly. She had sat down on the box and now she moved a little forward so that her feet rested on the ground. The sawdust that covered it gave off a clean, pleasant smell. It was striped by the long pear-shaped shadow of the single electric light bulb that dangled naked from the ceiling.

The pale shadow was swallowed up by a blacker one. She looked up at the flimsy wall that separated the box from Red's stall. A nail had come loose; one of the boards was sagging. Red had thrust his nose through the aperture and was pushing at the board. It flopped up and down, creaking. Red's whole head appeared in the opening. His dark eyes shone. He tossed his head up and down lightly. She got up and, laying the palm of her hand flat against his nose, pushed his head back. "Bad boy," she said, "you're a bad boy."

Mr. Shannon had come up behind her. He looked about for a tool and when he could find none slipped the board back into its place and hammered on it with his fist. "I'll have that fixed tomorrow," he said.

She had stepped out of the box and was looking over the gate into the shining eyes. "He seems sort of nervous to-night," she said.

"The crowd bothers him," he said. "They just have to get used to that. He's a lot better'n he was yesterday. I had to water him myself yesterday. That boy couldn't manage him."

"Did he cut up?"

"Reared a little." He tested the fastening of the gate. "He'll be all right, soon as we're gone," he said. "Stallions always got to be nibbling something . . . Miss Willy, what say we go and get something to eat?"

"I saw Elsie Manigault in one of the boxes a little while ago," she told him as they walked down the lane.

"Marble didn't come," he said. "I reckon he knew that filly of theirs wasn't going to show up so well. He didn't much want to show her, but Mrs. Manigault was set on it."

"Didn't she win?" Willy asked. "I didn't even watch the filly class. But I thought she was sure to win."

"It was a horse from Fayetteville," he said. He laughed. "I'd liked to a seen Mrs. Manigault's face when they give Red the ribbon."

They were on the mid-way. At this time of night it was crowded with people. Mr. Shannon suddenly stopped and, bowing slightly, offered her his arm. She laid her hand on it. He turned so that he walked a little ahead of her and made a way for them through the crowd. He paused before the tent which housed "Fatima, Queen of Fortune Tellers." "Want to go in?" he asked. She shook her head, smiling. They walked on.

After a little he stopped again and, getting out his handkerchief, mopped his forehead. "It was along here," he said, "that place Joe told me about was along here somewhere."

They crossed the street and joined a line of people standing in front of a pavilion which advertised "Sizzling Steaks —Short Orders." Inside the tent customers sat on stools at a shining counter, back of which steaks were being cooked on electric grills. Mr. Shannon led her past the counter into an adjacent tent where people sat at tables covered with red-and-white-checked cloths.

He sat down with a sigh. "This is the place," he said. A negro boy was at his elbow. "You got any draught beer?" Mr. Shannon asked.

"Sure have, Cap'm," the boy said.

Mr. Shannon looked at Willy. "Miss Willy, you like draught beer?"

Willy laughed nervously. "I don't know," she said, "I never tasted beer. Yes, I did, once. Jack had some in the refrigerator and I tasted it. I thought it tasted bitter."

Mr. Shannon laughed. "Ladies don't usually like beer," he said. "How about some lemonade?"

"Yes," Willy said, "I'd like some lemonade."

"Ain't got no more lemons," the boy said sadly.

"Well, orangeade, then," Willy said. "Have you got any oranges?"

"And a steak?" Mr. Shannon said. "I was too nervous to eat much dinner tonight. I'm going to have me a good steak. You have one?"

She shook her head.

"Got some nice chicken sandwiches," the boy said.

"I'll take a chicken sandwich," she told him, "and a glass of orangeade."

The steak, flanked by a mound of cole slaw and thick, brown slices of fried potatoes, was set before Mr. Shannon. He did not speak while he was eating. Willy nibbled her sandwich and drank her orangeade, gazing about her. The place had seemed crowded when they entered but more people were coming in. A slim, middle-aged woman in black came past, attended by three men. She stood for several minutes, steadying herself with a gloved hand resting lightly on the back of Willy's chair, while her cavaliers secured the last table left in the far corner. The perfume she used hung in the air for several minutes after she had gone. Outside it was still cool, but this air was warm and weighted with sounds and scents. From the four corners of the tent powerful, shaded electric light bulbs cast a glow on the flushed faces, on the red and white checks of the table cloths. Glass steins of beer stood on every table. Some people had bowls of ice before them and empty glasses into which they poured whiskey from bottles which they took from their pockets. The bottles, Willy noted, were never allowed to stand up straight after the whiskey had been poured but were laid down flat on the table or returned to the men's pockets. When she had first come in she had felt uneasy. She had never been in a speakeasy before. She had imagined that such places, in this part of the world, at least, were patronized only by men. But there were women here, too, respectable women, most of them better dressed than she herself was. A young girl at a near-by table had two beaux. She divided her time scrupulously between them, tilting her tawny head to flash a wide smile at one, or turning and with chin cupped in hand gazing dreamily at the table-cloth while the other spoke in a low voice.

Willy stared covertly at the girl, at the cloud of hair shading the brow, at the dropped lids which shone a little as if touched with some emollient, at the fashionably hollow cheeks, and tried to imagine what the man was saying to her. She was filled with lonely, impotent anger. She looked away. The woman in black was telling a story. Her three escorts fixed her face with bright, amused eyes. When she had finished, one of them put his arm about her shoulders in a quick, affectionate squeeze. "Lucy, you ought to be ashamed of yourself!" A man at the next table turned as if he thought he had been addressed and gave him a friendly smile. The group at each table seemed self-contained but all these people knew each other, could have called back and forth if they had been so minded. And they were not strangers to her. In a sense they were her lifelong companions. All her life she had stood, alone, listening to the sounds of their merriment float out from behind a closed door. But why? she asked herself. Why was it always like that? How am I different from them?

Mr. Shannon had finished his steak and was ordering another stein of beer. "You won't change your mind, Miss Willy?" he asked.

"Yes," she said suddenly, "I'd like some beer."

The boy brought the beer in a mug frosted as white as a julep glass. She raised it to her lips. As she had expected it tasted bitter. She took a long swallow and set the mug down.

Mr. Shannon leaned forward. His pale eyes studied her face. "Miss Willy," he said, "how'd you like to take Red to Shelbyville?"

"We're going to, aren't we?" she asked. 'Next year."

"He's eligible now," he said quietly. "He's won in the All Age. The entries don't close till Saturday . . . If I's to get a letter off tonight. . ."

She lifted the stein and took another draught of the beer, then set the stein down, swallowing quickly, to get the bitter taste out of her mouth. "All right," she said. She stood up. "Let's go back to the hotel now and get the letter off."

Mr. Shannon laughed. "That's the right spirit," he said.

She had already started on. She was in the outer tent before she realized that he had stayed behind to pay the bill. She felt conspicuous, standing alone. She went outside and stood in the street. The lights were not so brilliant now. Many of the shows had closed for the night. A little way down the street a group of people stood before a brightly lighted, pink-and-green-striped tent. A barker shouted: "Come and see the couple being married in the block of ice!"

"The Grand Championship," she thought, "the Grand Championship of the World!"

Mr. Shannon was beside her, his hand cupped under her elbow. They moved forward. He hesitated as they came abreast of the striped tent and looked up at the sign.

"Married in a block of ice!" he said. "I wouldn't like that much, now, would you, Miss Willy?"

"I don't think they ought to allow it," she said and would have passed on but his hand compelled her to turn with the crowd. "Let's just go in and see 'em," he said, "We been so taken up we hardly seen anything yet." His voice boomed out, suddenly genial. "Why, we hardly been to the Fair yet!"

They passed, with the crowd, through the gate. In the centre of the circular tent whose walls were ornamented at intervals with huge, pink roses, the man and woman lay, in a glass case that was enclosed in a long block of ice. One saw the groom first. His black hair, parted on the side, plastered his bony skull slickly. He had a cape jessamine in his button-hole. Willy looked down at the sawdust under her feet. It was perfectly dry. "The ice doesn't melt," she thought. "But of course. They drain it off. Under the case somewhere." Mr. Shannon's hand, under her arm, propelled her faster. "Let's look at the little lady," he said.

The coarse, white veil was long, to hide the scantiness of the white satin robe. In its voluminous folds the yellow-haired bride lay as unstirring as one of those nosegays which florists at Valentine time enclose in stiff, white lace paper. Under half-closed lids the blue glare of her eyes was directed at the opposite wall. Her full, rouged lips parted to reveal small, imperfectly formed teeth. A tiny pustule on the upper lip showed yellowish in the harsh light. The hand which clasped the bouquet of gladioli and ferns to the breast was plump, the skin reddened and coarse in texture, the short fingers ending in sharply curved nails, ornamented with blood-red discs.

"I don't like to look at them," Willy said in a low voice, and wondered if they could hear anything through the glass.

"They got in there themselves," Mr. Shannon told her. "We didn't put them in there. They got in there of their own free will."

But he hurried with her to the turnstile where a loud speaker proclaimed that each day at ten, twelve, two and four o'clock the Reverend Albert Miles would perform the double-ring marriage ceremony.

"I knew a couple was married on horse-back once," Mr. Shannon said. "Walking horses at that. Seemed like a lot of foolishness to me."

His hand closed firmly on the upper part of her arm. He propelled her more swiftly through the crowd. Willy looked back. The groom's face was visible through a gap between the heads. When they had come in he had been staring straight before him. But now in the pallid skull an eye moved, sending its dark, oblique glance out over the crowd. A pulse beat in the temple. The dark, straining, hostile eye met hers for an instant, then travelled on to rest on the tent wall. She turned her head quickly and plunged with the crowd out on to the street.

Mr. Shannon took out his handkerchief and wiped, first, his forehead, then the fingers that had been clasping her arm. 'Well, now, that wasn't much, was it?" he asked. "Seems like they had better side shows when I was a boy."

"When I was a child I could never bear to look at the freaks," she said faintly.

"A bearded lady or something like that," Mr. Shannon said, "Or a dog-faced boy. When I was a kid was a fine dog-faced boy used to come every fall to the Montgomery county stock show . . . But a couple laying up in a block of ice! That ain't nothing."

They came to an open space before a merry-go-round. "Now that's what I like," Mr. Shannon said. "My old daddy never could understand that. 'Why'n't you ride a tobacco stick?' he'd ask me. But he'd give me a quarter to ride on. Least, he'd turn my own money over to me. . . . I started working in the field when I was eight years old, Miss Willy. By the time I was fifteen I had me a hundred dollars saved up."

Willy's eyes followed the prancing wooden horses. "I like the white one," she said. She laughed merrily. "Look! His tail's set."

The machine was coming to a stop. The operator saw them and stepped quickly down to the lower platform. "Five cents a ride! Ride the blooded horses. Five cents a ride!" He caught a stirrup and dangled it at them, smiling invitingly.

Willy hung back. "We'd better go on to the hotel," she said. "We've got that letter to write."

They walked on. Mr. Shannon suddenly stopped and stood with his head tilted back, gazing. "Now, that's what I like!" he cried.

Willy gazed too. Light glittered on the steel base of the Ferris wheel but higher up the supporting framework was hardly visible. The gondolas, each curve outlined in pulsing light, might have climbed of their own accord into the dark blue sky, to sway as gently to and fro as lilies in a pond.

"Let's go up," Mr. Shannon said. "I won't hardly feel I been to the Fair till I been on the wheel."

He paid their fares. They stood, waiting, while the great wheel slowly elevated the cars which were already occupied. An empty car came to rest before them. They got in. The car mounted slowly. The swaying motion, which had been almost imperceptible at first, increased as the wheel settled to its journey. The night air was lively with the cries of those left behind. Willy looked down. The upturned faces had grown smaller. The cries which had been so various were all one cry. She felt herself growing a little dizzy and looked steadfastly before her. They had left the tallest of the buildings behind. In the dark sky only the radio tower showed its red and green lights. She turned her head so that it should be out of her range of vision, wanting no companion except the cool air and the dark.

Mr. Shannon moved closer to her. He pointed. "That's where we ate supper." She looked down. The double string of electric lights glowed brilliantly on each side of the midway but the tents looked flat and dark. The people were still in there, eating, drinking, calling out to each other. But the air, by this time, must be close. The eyes of the suppers would be glazed, their faces flushed from the beer and whiskey. She lifted her head and felt the air strike cold on her cheeks. Red Allen, by Roan Allen out of Miss Fancy. At Stud. Twenty-five dollars. They always charged at least twenty-five dollars for service. By January she would have enough money to buy a brood mare. There would be colts to raise. She would keep one. Just one colt. For her very own . . .

"Everything looks mighty small, don't it?" Mr. Shannon asked.

Theirs was now the topmost car. The swaying had increased. The gondolas wanted to leave the wheel and rocket down on to the dark plain. She felt her stomach draw inward sharply and turned her head agitatedly to the right. Mr. Shannon put his arm about her shoulders. "Don't look down," he said. "Don't look down now."

Willy sat rigid. The slight nausea passed. The car had

begun its descent. The lights below were brighter. The plump woman in the blue dress still stood with her face upturned, her eyes, blue and hard as marbles, glittering in the light. Somewhere behind her a child was crying. A black-haired man in overalls, who had been leaning against a telephone pole, suddenly straightened up and, taking a red handkerchief out of his pocket, began to wave it, laughing frantically.

He will take his arm away in a minute, she thought. It would be better if I didn't sit up so straight. Men always put their arms around girls on the Ferris wheel. She tried to relax her muscles. Mr. Shannon responded by leaning closer and tightening his arm about her shoulders. She could hear his hoarse breathing. The heavy body was warm against hers. He had actually moved over a little on the seat. His clasp was no longer impersonal but intimate. His fingers moved, gently caressing the back of her neck. The car was coming to a stop. He drew her close in a swift, half embrace, then released her.

The man in overalls, disregarding the operator's warning, jumped up on the platform. From the gondola behind a baby was handed down to him. Willy and Mr. Shannon descended and walked out into the crowd.

"We might as well go back to the hotel, I reckon," Mr. Shannon said. "Got to get that letter off."

Willy, walking with her head slightly bent, did not answer. She could not look up. The blood that had rushed to her cheeks had not subsided. She could feel it throbbing almost angrily. And she had been seized by an uncontrollable trembling. "If I can just get back to the hotel," she thought, "I'll tell him the Ferris wheel made me sick. I'll get him to fill out the entry by himself."

They passed through the turnstile and walked across the common. A procession of automobiles held the road. They had to wait several minutes before they could cross. The sidewalks were crowded. Mr. Shannon led her up on to somebody's lawn. They moved forward swiftly, skirting the crowd.

In the hotel they got the entry blanks from the clerk and turned into the ladies' parlor. Mr. Shannon drew a chair out from before a little black desk. She shook her head and moved to the window. "You write it," she said. "You can do it better than I can."

"All right," he said.

But he did not sit down at the desk. He stood, holding the papers in his hand, then suddenly thrust them down into his pocket and took a step towards her. "Miss Willy . . ." he said.

The blood had ebbed from her cheeks. The trembling had ceased. She fixed her eyes on his face. "What is it, Mr. Shannon?"

He looked at the floor. His head turned slowly to one side. The muscles of his lips twitched. The lips themselves pursed, as if they found the words he was about to utter difficult of enunciation. He shook his head again, sharply, like a man beset by bees. "Miss Willy," he said, "you know what I mean!"

She continued to regard him calmly. "Is it something about Red?"

He came nearer. "It's not only the horse," he said, "but

it looks like we've got a lot in common, living neighbors and all." He thrust both hands into his coat pockets and balancing on his toes inclined his heavy body a little while his gray eyes held hers. "Miss Willy, I'd be a happy man if you'd marry me."

"I couldn't," Willy said quickly. "I—I wish I could. But I couldn't."

There was a silence. His gaze sharpened. "The old lady won't like it," he said. "I know that. But she won't live forever. You got to remember that, Miss Willy."

Her face was flaming. She held her hand up between them. "Please Mr. Shannon. It's not Mama . . ." She had started forward and now she was directly opposite him. Her hand, anguished, groping almost as if she were blind, reached out and for a second lingered on his. "I can't explain. But I couldn't . . . It's no use to discuss it."

He had turned as she moved and kept his eyes fixed earnestly on her face. When his look grew more intent she lifted her hand swiftly from his, pressed it against her throat. "Please, Mr. Shannon . . ." she said.

"I reckon not," he said heavily.

She was at the door. He took another step towards her. "Miss Willy . . ." She turned around. "I'm not in the notion of marrying, just to be marrying," he said. "You realize that, don't you?"

"Yes," she said faintly.

"I just wanted you to know that," he said, "if you were ever to change your mind . . ."

She shook her head. Her hand went out to him in another, impotent, distracted gesture. He suddenly laughed.

His own hand, going down into his pocket, brought up the entry blanks. "You go on upstairs," he said. "I'll 'tend to these."

She left the parlor and crossed the lobby. In the elevator the boy was dozing. She tiptoed past him and walked up the single flight of stairs to her room.

She turned on the light, took off her hat and jacket and sat down on the bed. The dresser was directly opposite. The woman framed in the oval mirror sat unnaturally straight on the side of the bed. Her breast rose and fell with her hard breaths and the eyes in the darkly flushed face glittered. Willy looked into the eyes for a second, then, with an hysterical laugh, fell to one side, burying her face in the counterpane. The pillow, she thought, as her face touched the hard, smooth surface. I left it on the stands. But what does it matter? She went to the window and stood looking down into the street, empty now of all save a few late strollers, then, slowly, languidly, undressed and got into bed.

The bed faced the window. She lay quiet, staring at the sill washed white by light from the street below. In her imagination the words that had been exchanged in the parlor still sounded, louder than at the moment of utterance.

I'd be a happy man if you'd marry me. He had kept himself from moving forward by thrusting his hands deep into his pockets but all the while he was speaking his body had slowly inclined towards hers. And his gray eyes had had a strange shine. That was because they had been fixed so intently on her face. People look at each other a hundred times a day but it is seldom that any one regards you intently. It is so seldom that any one has anything to say that cannot be said with words, that must be conveyed by the eyes. By the eyes, if you are not close enough to touch each other. It was not accidental, there, on the Ferris wheel . . . *A happy man if you'd marry me* . . . He is a gentleman—I don't care what they say—he is a gentleman. If I had moved, made the least sign, he would have taken his arm away or kept it there so lightly that I would not have felt it. But I did not move. I did not make any sign. So he moved closer. He must have felt me trembling when his hand touched my neck. That was why he hugged me to him just before we got off. And in the parlor, when we first came in, he thought that I would understand, that he wouldn't have to put it into words. But I stood there, pretending I didn't understand him. That was when his look changed. But he did not go away from me all at once. He had to be certain, so he kept on looking at me. Then he was convinced. He turned away. I felt as if I were about to die . . . Don't go on waiting for me. You are too good, too kind. Go on and marry somebody else. I could not say it but I stretched out my hand. It was all there, throbbing in my fingertips. But he had stopped looking at me. He was already gone away . . .

A good thing. You never had any idea of marrying Quent Shannon. It would be unkind to trifle with his affections.

. . . When he first put his arm around me, there, on the Ferris wheel it was just to hold me in the seat. Men always put their arms around girls on the Ferris wheel. And it was just that at first, his arm steady and light about my shoulders, and then his hand came up and he waited a second and when I did not say no his hand slipped down inside my collar and his hand moving so softly and the close, warm feeling and only us two, high up in the air, where nobody else could come . . . It is not too late. I could change my mind . . . Miss Lewis. Miss Wilhelmina Gates Lewis. No, Mrs. Shannon. . . .

Quentin Durward Shannon. His father was reading Quentin Durward the night he was born. Quent can read too. He got as far as the eighth grade . . .

She had been lying on her back, still breathing short, hard breaths and now her mouth felt dry. She ran her tongue over her lips and turned over on her side and lay looking out of the window.

The knock came again. She sat up in bed, staring. She realized now that this was the second time it had come. But he ought not to come to my room, she thought. Somebody will come by, somebody will see him. She leaped from the bed and, catching up the robe she had flung on a chair, drew it on, and, pausing to fasten it with trembling fingers, went to the door.

"Who is it?" she asked in a low voice.

"Mrs. Manigault," a voice said, "Elsie Manigault . . . Willy, you aren't in bed?"

Willy slowly opened the door. Elsie Manigault entered. She cried out when she saw Willy in her dressing gown. "Oh, my dear, I'm so sorry! The clerk told me you'd just gone up. I thought you couldn't be asleep . . . I've been trying to get you all day . . ."

"It's all right," Willy said. "I'd just got in bed. I wasn't asleep."

She would have turned on the light but her visitor with

a gesture dissuaded her. "I can't stay a minute," Elsie Manigault said but she let Willy remove a heap of underclothing and sat down in the only chair. Willy sat down on the edge of the bed, drawing her robe a little more closely about her.

"When did you get here?" she asked after a pause in which her visitor had not spoken.

"This afternoon. I came straight to the hotel but you'd gone to the ring. Then I dined with the Powells and by the time I got back here you were gone again . . ."

"Is everything all right at home?" Willy asked.

Mrs. Manigault made a vague gesture with her gray gloved hand. "Really . . . Catherine has treated us shamefully. She hasn't been to see us since you left."

"I expect she's pretty busy," Willy said.

Finding that she was trembling again she swung her feet up on the bed and slipped down inside the covers, drawing them up almost to her chin. "I'm sort of tired," she said, "sitting in that grandstand all day . . . Cousin Elsie, I was sorry I didn't see your filly show. She won second place, didn't she?"

"Third," Mrs. Manigault said. "A little bag of bones from Fayette county took second. Really, these judges! . . ." She took her cigarette case out and after extending it perfunctorily towards Willy, lit a cigarette. "I was delighted to hear that Red won in the stallion class," she said. "Joe was amazed. He didn't think he ought to be shown this year."

"Catherine told me," Willy said. "He didn't think he was filled out enough, she said."

"He was a little worried about his manners, too. You

know, Willy, if you are going to keep on showing him you ought to have some competent advice . . ."

"If I keep on showing him?" Willy repeated. "Why, he's just started, Cousin Elsie."

Mrs. Manigault laughed. "I realize that. But you might not keep on showing him yourself. You might sell him. Had you ever thought of that?"

Willy stared at her. In the half-light from the transom Elsie Manigault's face looked drawn. Her vivacity masked a nervous intentness. She is talking as if we were friends, Willy thought, as if she came often to Swan Quarter. But we have never been friends. Why has she come here now? And then she reminded herself that Mrs. Manigault was always polite, even gracious to everybody. But it was the abstracted, impersonal graciousness of careful breeding. Her graciousness tonight was a little exaggerated. She has only come because she wants something, Willy thought. *She wants Red!* He is all I have and she knows it and wants to take him away from me. But she cannot have him. If I can't have anything else I'll have him. I won't sell him. No matter what they offer. I won't sell him.

Aloud she said, "I'm never going to sell Red, Cousin Elsie."

"Never?" Mrs. Manigault asked lightly. "In that case you really ought to get some competent advice. Joe will be delighted to help you in any way he can, of course. But it's tough going, for a woman alone."

"Mr. Shannon's a good man with horses," Willy said. "At least that's what everybody around here has always thought."

"He used to be a jockey, didn't he?"

"Yes."

"Well, times have changed . . . You really mean that, Willy?" she asked suddenly. "You don't want to sell?" She stood up. She approached the bed. "I'll give you ten thousand dollars," she said in a low voice. She made a quick, controlled gesture towards the desk in the corner. "I'll sit down now and write the cheque."

Willy shook her head. "I'm not going to sell him," she repeated.

"It's no use asking any more," Mrs. Manigault said. "I won't give it." Her voice rose, cracked a little. "That's a ridiculous price for any two-year-old. But I want to see what we can do with him. It worries me. I'm afraid that Shannon will ruin him. After all, he was our colt first . . ."

Willy sat up in bed, her arms clasped about her knees. "I've had all that out with myself," she said steadily. "You could have kept him. But you didn't. You gave him to me."

Their eyes met and held each other for a moment. Mrs. Manigault looked away. She laughed nervously. "Oh, of *course!* I wasn't disputing your ownership. I just wanted you to know how I felt. Joe and I both want to do everything we can to help." She picked her glove up from where it had fallen on the floor, straightened it out with a little flip and drew it on. She opened the door and stepped out into the hall. "Good night, Willy," she said.

"Good night," Willy said and when the door had closed behind Mrs. Manigault she turned over on her side again, pillowing her cheek on her hand. While she was talking with Mrs. Manigault she had been assailed by a deathly weariness. At that moment she wanted nothing but sleep. But now that her visitor had gone she could not sleep but lay staring with bright eyes at the lighted transom. Elsie Manigault had been trembling too, so much that she could hardly control her voice when she said good night. It was not disappointment at not being able to buy the stallion that put that look of unutterable fatigue in her eyes. It was some other disappointment, some other never-to-be-explained loss, something she had always wanted and never got and now feared that she never would get. Nobody gets what they want, Willy thought, and lay watching the events of the day recede, losing little by little their fiery brightness until they seemed like something that had happened a long time ago, to be remembered, but only at intervals and that more and more rarely, until finally the recollection disintegrates, like paper which, when it has lain long enough, turns brittle, so that, finding when we open the casket only the brown, dry fragments, we forget what was written on the fair page and do not visit the room that harbored it any more.

XIX

WHEN CHAPMAN BOARDED his train at midnight his berth was made up and he went at once to bed. But he lay sleepless, thinking of the abrupt leave he had taken of his duties at the university and of the trip abroad he would shortly make. Two weeks ago he had written to a friend in the State Department, telling him that he was going to Italy, that while he was there he intended to talk with a certain nobleman, still in office, though known to be inimical to Fascism. If the State Department were agreeable he would place what information he was able to get at its service when he returned. Hampton Means had telephoned him that night: "When can you leave?" "The *Savoia* sails on the fourteenth," Chapman told him. "I can make that." He had gone down to Washington and had talked with Means and with the under-secretary and was about to return to New York when he realized that his boat would not sail for ten days. There are occasions when we seem to have outrun time, to be standing, waiting for an intolerably slow companion to catch up with us. The night that holds back a lovers' meeting will, in memory, cover a span of years; we can face a lifetime with more equanimity than the half hour that intervenes before a fateful encounter. It seemed impossible to Chapman that he should return to New York. He called the Union station, found that a reservation had just been cancelled, and boarded a south-bound train that night. He had had an hour free before his train came in but some instinct kept him from communicating with Catherine. But he would have to let her know he was coming if he was to be met at the train. He decided that he would telegraph her as soon as he woke in the morning.

He woke at eight o'clock. The train was passing, at full speed, through a country town. On a bench before a white-columned porch loafers were already sunning themselves. From the square a red-clay road spun out, parallel with the railroad tracks. Two boys were driving a herd of cows along it.

He went into the washroom, dressed and entered the diner. It was crowded but he found a table that had only one occupant, a black-haired young man about thirty years of age. A waiter brought a platter of ham and eggs and a pot of coffee. The young man lifted the newspaper he had had spread out before him, folded it and handed it to Chapman. "I'm through with it," he said.

"Thanks," Chapman said and, holding the folded paper in his hand, studied the menu. "I'll have ham and eggs, too," he told the waiter. "No orange juice."

He looked at the newspaper: *Flying low through scudding clouds, German bombers returned to the attack for the thirteenth successive night, after a day of comparative quiet . . .* But the RAF Bombers have set Havre afire, he thought . . . *U S May Use British World Bases . . .*

He raised his eyes from the newspaper. The red road had slipped from the right to the left side of the tracks. The cows, changed miraculously from Holsteins to Jerseys, were being turned through down bars into a pasture. One of the boys had suffered no metamorphosis. The other had become a young man, hatless, his blue jeans girded low on his lean hips. He felt himself observed and turned and meeting the fleeting gaze, winked and raised his thumb to his nose.

The waiter brought Chapman's order. Chapman wrote his telegram and handed it to him. His companion was rising. "Don't you want your paper?" Chapman asked. The young man smiled, evidently disposed to be friendly. "Help yourself. By Jove!" he added, looking down at Chapman's plate, "I believe you got more ham than I did."

"Naw suh, naw suh!" the hovering waiter said quickly.

Chapman laughed, lifting a forkful of eggs to his mouth. "I always eat ham and eggs on the train," he said. "Can't stand 'em any other time."

The other laughed too. *"You may talk of gin and beer!"* he said. He lingered a second, seemed on the verge of saying something else, then abruptly laid a quarter beside his plate and departed.

Chapman, taking his first draught of hot coffee, looked after him. An odd remark, that, about the gin and beer, to come from a fellow with a jib cut in that fashion. The young man, tall, well set up, with unusually broad shoulders, looked as if he had been on Wall Street but still kept up the athletic sport in which he had distinguished himself in college. But the glance of the dark blue eyes was at once sharp and speculative. There were slight pouches under the eyes, and the cheeks, freshly shaved but still faintly stubbled, showed the network of fine wrinkles that tell of long hours spent at a desk. An editor, Chapman thought, *Life* or *Fortune.*

He ate his ham and eggs, drank a second cup of coffee and went into the club car to smoke. It was crowded. There was only one vacant seat. As he sat down he saw that the occupant of the next seat was the young man to whom he had talked in the dining car.

He exchanged greetings with him, lent him a match and turning his chair slightly away, gazed out of the window. He had had a letter from Edith just before he left: her second letter since he had last seen her. She wanted, she said, at least to have a talk with him. But if he could help it he would not see her again. There was nothing he could say to her. He had behaved badly, going to her apartment drunk, leaving her the next morning without a word of explanation or apology, and now he was on his way to see another woman, to whom he had behaved even worse. If he went back to Edith she would weep a little and finally tell him that nothing he had done made any difference. But Catherine, even if she loved him—he did not know whether she ever *had* loved him—might stand up as straight as that poplar outside the window and tell him to go to hell. He had always known that she was capable of hate. If she had come to hate him she would say things that he would remember as long as he lived. He thought as he gazed out of the window that he was a fool to go down there.

It was a late September landscape. A meadow, cleft by a brook, ran in yellow and russet waves of sedge grass up to a wood. There was a drift of fallen maple leaves on the wood's fringe but the tall, straight trees, the hickories, the oaks, the poplars, still pushed green boughs against the sky and few of the leaves had turned, except where a poplar showed amid its dark green a sudden flare of pure yellow.

He watched the sunlight creep out of the woods and run, glistening, on freshly turned red earth. In the north, this time of year, the leaves have all turned, he thought, and the colors are sharper . . . This is a softer landscape. Not wild, like Maine; neither park nor cemetery, like Connecticut, but a landscape dominated by man, tutored to his needs. And it seemed to him that the landscape was alive, a great beast, with flowing, sinuous limbs that disposed themselves in various attitudes, a gentle beast that stood or gazed or marched docilely, drawing behind it the great wain loaded with harvest fruits.

He told himself that it is not from Theocritus, from Bion, from Bacchylides that you get the feeling of the Greek landscape, but from Hesiod. And the picture comes to you, not from the shepherd, who, straying into the woods or mountains, meets with a god or goddess, but from the glimpses of homely country living that Hesiod affords. The gray-eyed Athena, visiting Ulysses' palace, wraps herself in a cloak of rough, home-spun weave. Ulysses ploughs his own fields.

The dark young man had turned so that he now faced Chapman. He spoke, smiling. "Pretty nice out there."

He pronounced the word "out" like a Virginian. Chapman eyed him curiously. "This your country?" he asked.

The young man shook his head. "My people came from

Fauquier county, originally. But they've been settled in Kentucky for five generations. I'm from Paint Lick, Kentucky." His smile enclosed the name of the village with inverted commas. "Paint Lick, Kentucky," he repeated, as if relishing the sound of the words. "My name is Napier. Edmund Napier."

"That's the place where the Indians painted the great totemic animals on the rocks, isn't it?" Chapman asked.

Napier stared. "How did you know that?" he asked. "When I was a kid there was an old man used to tell me about those paintings, but there's nobody living now who has seen them."

"Oh, I read it in an old book," Chapman said. He laughed. "I teach history for my living."

He looked out of the window again, thinking of the country as it had been in those days. The pioneers had settled by those salt licks as the early Venetians had settled on the salt marshes of the Adriatic. Catherine's people had come into their region upon the heels of the pioneers and had settled upon a Revolutionary grant. To Irish John Lewis. No, to his sons. There had been two of them in the battle of King's Mountain."

"Where are you from?" Napier asked.

"My name's Chapman. I come from Ohio," he said, "but I always marvel at Southerners."

"Why?" Napier asked.

Chapman shook his head. "I'm damned if I know. I suppose it's because they always come from somewhere." He smiled suddenly. "And they have such distinguished ancestors."

Napier laughed. He looked out of the window reflectively. "My most distinguished ancestor was Edmund Ruffin, of Virginia," he said. "Know who he was?"

Chapman nodded, startled to hear the name of Ruffin on the lips of the young man.

"He was a pretty bright old boy," Napier said, "Edited *The Farmers' Register,* which in its day had quite a following. When the Civil War came he went to Charleston and joined the Palmetto Guards, though he was then over seventy. Fired the first and, you might say, the last shot at Fort Sumter . . . When he heard that Lee had surrendered he put a bullet through his head. Left a note saying it was all worse than he'd expected . . ."

Chapman was still eyeing the speaker. The face was one that he had often seen before, alert, disciplined, histrionic rather than contemplative. "I wouldn't want Ruffin for my grandfather," he said.

"Why?" Napier asked, amused.

"He was a great man," Chapman said. "There's a theory that if the germ plasm works too hard in one generation it takes a rest in the next." He heard his own words with surprise. It was the face, he thought. He saw it too often, whenever he stepped out on the street or entered a colleague's office. There had been a time when he could escape it by shutting his own front door. But of late it had invaded his nights. Night after night it pursued him down an unending corridor. The thing, he thought suddenly, was not to flee the face but to turn and shatter it. He leaned forward, looking straight into the opaque blue eyes, relishing the brutality of his own words.

"I don't expect you've got much on the ball," he said.

Napier laughed. "I don't know," he said, "I've been pretty lucky." He smiled back at Chapman, the frank smile of the man who is accustomed to finding himself liked wherever he goes. "I was at Eritrea when the Italians moved in. In a barber shop with four chairs. Haile Selassie's son-in-law walked in but said he didn't want a shave. He bought a bottle of Italian perfume."

"I don't know," Chapman said abstractedly. Years ago when he had been a young man he had kept propped before him one long afternoon in the basement of a Southern university library an out-size calf-bound volume: letters from Colonial and Revolutionary governors, to one another and to the Indian chiefs whom they sought to defraud. The letters to the chiefs, which, according to convention, began "Dear Friend and Brother," were meaningless, except for what could be read between the lines. But the other letters, from one statesman to another, when divested of formal compliment, always contained a kernel of fact. Governor Blount would or would not support a proposed measure. He considered a certain colonel a scoundrel and would have nothing to do with him. There had been a post-script to one of the letters: "Joseph Ray arrived here last night, having made his way through the Creek Nation, and bringing me your letter. It was much stained with his blood."

The trouble is not with this young man, Chapman thought. It is the words. They have grown light and anybody can pick them up. There are too many of these word-men and now they can fly and they cloud the skies over the oceans and a man on one continent has less chance of communicating with another than when the letter was brought by sail and carried in a leather breeches pocket five hundred miles through the wilderness . . . *Much stained with his blood . . . But there is no blood* . . . The boy before him suddenly stepped agilely on to the green, cushioned seat and from there into a plane, whose wings flashed silver as it ascended towards the sun. The plane ceased to climb, hung motionless in the thin air. The face grinned as it sped past, but the torso showed a more terrible grin, the wide lips of the disembowelling wound gaping to spray upon the rejected earth Klieg lights, television sets, bomb sights . . .

I am going crazy, Chapman thought. It is not so much the difficulties of my private life as pressure from without. No, the pressure comes from within. I dreamed the other night that I had swallowed the globe, and the continents, angry at being imprisoned, churned and groaned, rubbing their shores against each other. The pain was so exquisite as to be indescribable. Like child-birth, probably. At any rate it carried its own anodyne, for I have not remembered the dream until this moment . . . It has been coming on a long time, this insanity. This is the first attack. It is unfortunate that this fellow should witness it. He will make for his club the first night he gets back from Paint Lick and will run into somebody I know. It will be all over town. Catherine will have no trouble in divorcing me . . . But she would have no trouble in any case. You do not oppose a woman who wants to divorce you nowadays. It is the same as it is with many savage tribes. A few words spoken in the presence of witnesses will do the trick. In this case ten minutes in a judge's chambers . . .

He rose. "I'm going outside and get some air," he said.

The train was passing through a deep cut. He stood on the platform, smoking, and watching the red clay banks, fringed with pines, slide past. The landscape that, a few minutes ago, had disported itself outside the window, a great, companionable beast, now, gaunt and ravaged, fled the train. And the narrow shell itself hurtled to no purpose. In a few hours, having attained no goal, it would turn and speed north. And all these pines and gullies will turn, too, he thought, and run just as crazily in the other direction. Like this journey I am making. It cannot come to anything. I do not know why I am going.

He thought of the adultery that had estranged him from his wife. It was only recently that he had thought of it as adultery. At the time his affair with Edith had seemed a diversion, an excursion that anybody might be permitted to make. But he could fix the moment when it had become possible for him to make it. The night when they were out with Koenig and it had first occurred to him that Catherine might actually have taken Koenig for her lover. He had not thought of questioning her. It had seemed preposterous to him that he should, for a moment, appear in competition with the painter. His vanity, he supposed. It had smarted for days. But more than his vanity had been involved. In the moment when he had had those suspicions of her she had ceased to be herself, had become, for him at least, another woman. The woman with whom he had been in love, to whom he had been married for years, had disappeared, leaving a stranger in her place. He had regarded the stranger with aversion and had set out to win another woman. The pursuit had been short, the conquest, he thought wryly, not difficult, and it had not brought him much pleasure. Indeed, since his wife had gone from him he had felt a deeper depression than he had ever before known. There were times when, in an illusion that was part dream, part waking, he seemed to be suspended precariously over an immense pit. He half knew what lurked in its depths, but his concern was not with avoiding the descent; he rallied his febrile energies in order that he might experience the fall. His detachment from the scene was the ultimate horror. He strove to realize that he was falling, and could not. She was gone and since then he had been absent from himself. Was the sexual act surrounded with mystery because it was, in essence, magical? Did the woman who once truly received a man become the repository of his real being and thenceforward, witch-like, carry it with her wherever she went?

The truth is that I have no character, he thought. He thought of a colleague in the university, Wilkins of the English department. "Wilkins is always on the right side," Dick Reynolds had said of him once. He was not expressing any great admiration for Wilkins, who was not a man of much intellect or deep convictions. He meant that, held upright by his instincts, his prejudices, Wilkins was never swayed by personal considerations or university politics, and was always ready to champion the under-dog. But I have no prejudices, Chapman thought, no instincts, no convictions that are readily translatable into action. Wilkins is himself, the way a dog is a dog or stinkweed stinkweed or dog-fennel dog-fennel. The only times in my life when I have ever really existed have been when I was scourged by some idea. You are a well of loneliness, my boy, he told himself, casting the burning butt of his cigarette from him. No, a cave that at present has no bats flying in it . . .

He went back into the car. Three men were looking for a fourth at bridge. He joined them and played until lunch time. After lunch he read for an hour, then dozed in his chair. When he woke, the correspondent's seat was occupied by a young naval ensign. A magazine that he had been balancing on his knee fell over on Chapman's seat. "I'm sorry, sir," the boy said and bent to retrieve it. The effort sent red surging up under his close-fitting collar band. Even the healthy flesh over his cheekbones was suffused with red. Chapman smiled at him. "Not at all," he said, "Where are you stationed?" he asked and smiled again, "if it's not a military secret."

"I'm on leave," the ensign said, "but I've been out on the Yangtze Patrol."

"Good Lord!" Chapman said. "That must have been exciting."

The young officer nodded. "It's nice work. I've been there since last October."

"Why, you must have seen the *Panay* go down," Chapman said.

"I didn't see her but I talked to some of the guys that got off," the boy said.

"You think the Japs meant to sink her?"

"The flag was painted the width of her deck," the boy

said. "We could see 'em machine-gunning some of 'em when they tried to leave the ship."

"Deliberately? You saw that?"

"Sure. They turned the guns on 'em when they tried to get on shore, too."

"How long have you been in the navy?" Chapman asked.

"Five years."

"You're young for an officer. Did you go to Annapolis?"

The boy grinned. "I came up through the hawse pipe," he said. "Left home when I was sixteen. Enlisted in New York. Sailed on the *Lexington* as gunner's mate. Got a Presidential appointment to Annapolis two years ago."

"You think the Japs want war?" Chapman asked suddenly.

"Well, I don't know. There's a lot of Jap ships anchored near us. There's a lot of reviewing going on all the time. Don't look like they intend to stand idle."

"The Japs are really mean fighters, aren't they?"

The young officer's straight brows drew together in a frown. Under them the sunny eyes remained unclouded. "Mean as garbroth," he said.

Chapman realized suddenly that he spoke with a pronounced Southern drawl. And he had just used an expression that Catherine often used. "Mean as garbroth." (She pronounced it "gyarbroth.") "Garbroth? Why, it must mean broth—some kind of broth that is awfully mean . . ."

Phrases of this kind, surviving like the outcroppings that tell the geologist that the rock before him is made up of strata from three or four ages, or like the ruin that in one of Poussin's landscapes, reminds the gazing shepherds and shepherdesses that other shepherds and shepherdesses have sported in these same shades, had always delighted him. The person who uttered them, speaking, even if unconsciously, with the voice of the past, became more interesting than he would otherwise have been.

There had been a cousin of Catherine's, an old man, with sparse gray hair curling on a massive skull whom, if you visited Swan Quarter for as much as a week, you were always taken to see. In the summer his one-story, six-room house was too hot for human habitation, so he sat out under the trees, books and magazines piled about him, hens scratching a dusty circle around his rocking chair. He read Josephus and Plutarch and Herodotus and Taine and Gibbon but he liked to talk better than to read; he seldom found a listener who was capable of following his discourse. When Chapman was brought to call upon him he would greet him almost curtly and begin talking, usually about the Civil War. He knew the manœuvres of all the major battles in the west and particularly delighted in describing the siege of Fort Donelson, whose guns he had heard roaring one summer afternoon when he was fourteen years old. An insufferable old bore, when he wasn't talking about the battles, but Chapman, at times, had found himself attending the monologue with pleasure, not for its substance but for the sound of the voice, which, with its mellow tone, its old-fashioned turns of speech, seemed in the burning heat, under the summer trees, to set echoes chiming from a vanished, now almost forgotten world.

"I'm going to get new orders when my leave's up," the ensign said. "I sure hope they keep me in the Pacific."

"I can see that you would," Chapman said absently. He tried to imagine the boy manning a gun, standing with spread legs on a windy deck, shouting an order, but his imagination revolted at the picture. It seemed preposterous that such responsible work should be entrusted to a youth whose body was not as yet mature. He examined the boy's face. Nature had not accomplished her work but had merely indicated what the face would be like when she had finished. The flesh that clothed the long jaw was of the color and transparency of a child's flesh, and a child's fresh color showed in two bright spots over the cheekbones.

He reminded himself that in the eighteenth century boys of sixteen often held the rank of major or as captains commanded ships. This boy was twenty-two, twenty-three at the most. We will have war very soon, he thought. I did not realize it before. That is because I am a fool. But he knows it. They always know it. And there does not need to be any communication between them. They do not pass the word along, one to the other, but one day, all of a sudden, in the Kentucky mountains, in Newark, New Jersey, in California, in the Middle West they look up and know that it is time and stop whatever they are doing and come crowding down to where other men, old men like me, or women, are already caulking the ships. And we look at them and our hearts smite us and we throw flowers in their path and garland all the ships. But they cannot hear our cries and, embarking, happy in the certainty of their goal, turn their calm eyes on us and our hearts shrivel and the beaches are desolate and cold. It is not every generation, he thought wearily, for the human spirit could not sustain the journey. But it is always the fairest. I was too young for the last war and am too old for this war. But I wish I was one of them, for it is something, in this life, for a man to know where he is going, even if the appointment is with the minotaur.

The conductor came through, calling "Carthage!" The young officer jumped up, pulled his bag down and joined the procession in the aisle. His head turned, the impassive look gone from his face, he gazed eagerly out of the window. As handsome a boy as one would find in a day's journey. In a moment some woman would be clinging about his neck. She ought to have married somebody like that, Chapman thought, and he, too, stooped and gazed through the dusty window. At last he caught sight of his wife, standing alone on the platform, holding Heros on a leash. In dust-stained riding clothes, a red cap such as deer hunters sometimes wear in the north woods pressed down on her fair hair, she seemed to him the embodiment of all that was desirable. The passengers moved forward more briskly. He followed them out onto the platform, went down the steps. The porter was holding his bag. He took it from him and hurried towards her.

XX

HE PUT HIS BAG DOWN, set his hands on her shoulders and bent and kissed her on the lips. Heros was jumping about his legs. He picked the dog up. Heros reared against him, whining and licking his face, then jumped from his arms to the ground. The leash had been jerked from Catherine's hand. Chapman stooped and retrieved it.

"Is that all the luggage you've got?" she asked and when he nodded, led the way across the street to where she had parked the roadster.

Heros was still jumping frantically and whining as they seated themselves. Chapman held him on his lap and tried, with caresses, to quiet him.

"I was afraid he wouldn't remember me," he said.

She laughed. "I told him this morning that old Shack was coming," she said, using a nickname, which according to family fiction, was the name the dachshund called his master. "He's been just mad ever since."

"Really?" he asked and his big hand closed on the sleek, squirming body, pressing it still closer. "I wouldn't have thought he'd remember me by name. I would have thought it was by smell."

She opened her lips, seemed about to say something, then closed them. After an almost imperceptible pause she spoke. "I'm afraid sleeping beside the radiator has dulled his sense of smell. I'm afraid he just has to depend on the spoken word."

Chapman did not answer. When she had first opened her lips it had been to "speak" for Heros. He was about to make one of those spirited, derogatory remarks he so often made. But such badinage did not seem to her appropriate, now that they were no longer a united family. She had quickly substituted another remark, one that might have been made to a stranger.

Shall we have it out now? he asked himself. I can make her stop the car and draw up alongside the road. We can have our talk now, and then, if she doesn't want me to go on home with her, I can turn around. I can go back to New York. But there would still be ten days to kill before his boat sailed. He could not endure the empty apartment. It would be senseless and expensive to stay at an hotel. No, he thought, I've come, and I'll stay. We have got to understand each other. No matter how it turns out.

The town was behind them and now they left the highway and turned off on a country road. They descended a little hill. The car rattled across a wooden bridge. The foliage that overhung the shrunken stream, touched only here and there with autumnal color, was everywhere lightly filmed with dust. "It's been pretty dry, hasn't it?" he asked.

"Not until just lately," she said. "When I first came we had plenty of rain."

They left the bridge and started up the slope. Chapman's nostrils widened as he caught sight of a gaunt, wooden building perched on its summit. The familiar, nostalgic smell that he had been trying to identify was tobacco smoke. It curled in blue wisps from the small openings under the eaves of the barn. He realized now that the whole countryside was hazy with it.

"They're firing tobacco already!" he said.

"They started a week ago," she said and turned into another, smaller road that he did not remember.

"Is this the way we go?" he asked.

She turned her head and for the first time looked directly into his eyes. "We have to go by the Manigaults," she said. "That is, if you'd like some tea."

He would have preferred to go directly to Swan Quarter. But he did not know what he would say to her. He did not know that she would listen to anything he said. Perhaps it would be as well to postpone their talk another hour. He was not yet certain what her mood was. She had been cool enough when she greeted him. But he had expected that.

"I see you've got a horse," he said.

"Willy has a wonderful horse but he's away at the Fair now . . . The Manigaults lend me a horse whenever I want one."

"Have you been riding today?"

"Yes," she said.

"How is the family?" he asked.

"Willy's at the Fair, showing her horse. . . . Cousin Daphne is with us . . . Mammy," she said and arched her upper lip so that her teeth showed briefly, "Mammy is not exactly *with* us. She had a stroke the other day."

"A *stroke!*" he said, startled. "Was it very bad?"

She faintly compressed her lips. "I don't know. I never

had experience with any other stroke. We came in and found her lying on the floor, all covered with blood . . . She's up and about now. She seems quite well, physically. But she's out of her mind."

He remembered the old lady as almost belligerently clear-headed. He found it hard to visualize her in such a condition. "Do you mean she raves?"

"It isn't raving," she said, "It's just talk . . . I never heard anything like it before. She doesn't know who she is or where she is. She's always trying to find out . . ." She lifted a hand from the wheel, made again the faint grimace. "She's like a baby. She can't control herself, in any way."

"Good Lord!" he said. "You must have had a tough time, with Willy away! Does she know?"

"I thought it was better not to tell her. She has a pretty tough time when she's there."

Her compressed lips, her veiled look directed at the road ahead, seemed to reject the sympathy that had been implied in his tone. He fell silent, watching countryside that seemed to him surprisingly unchanged in the five years since he had seen it.

Heros had slid from his knee and was now curled up quietly between them, as he had lain so many times in the past. Chapman thought of the time when his wife had been pregnant. They had waited several years after they were married before they had even considered having a child and then, in 1928, she had become pregnant. There had been a fall on an icy pavement, an injury to her back that had resulted in a still-birth some months later. The doctor had advised that she should not become pregnant again for some months. A year had gone by. They had again considered having a baby but she had wanted to go abroad with him that summer and they had decided against it. It was two years later that they had bought Heros from a Long Island kennel, recognizing tacitly that although they had no children they must have something live and young about the house. It takes will to have children nowadays, he thought. I did not even have enough will for that. He took one of Heros' long, sleek ears in his hand and gently twisted it. The dog turned and fixed shining, inquisitive eyes on his face. "Whatever became of the Manigault boy?" Chapman asked.

"He's farming," she said. "He and Mrs. Manigault have divided the farm and each runs half."

"Mrs. Manigault!" he said. "Has she settled here?"

"Apparently. She raises horses."

"What does Tom raise?"

"Tobacco. Corn. Alfalfa. Herefords," she said curtly.

"You wouldn't have thought a boy brought up as he was could make a farmer," he said idly. "Is he a good farmer?"

"Mrs. Manigault thinks not. But then he doesn't approve of her farming, either."

They turned off the road into a driveway that seemed unfamiliar. Catherine had told him in one of the two letters she had written him since they had separated that Mrs. Manigault had razed the old house and erected a modern mansion in its place, but he was not prepared for this expanse of gleaming whiteness.

He said, "Is this Mount Vernon?"

She was shaking with laughter. "Roy Miller," she said. "I thought you'd like it."

They saw Mrs. Manigault standing in the open doorway. They composed their faces. Catherine parked the car on the gravelled drive. They descended and moved towards the house. He had felt that Catherine's long stay in the country must arouse remark and had braced himself against curious glances, slily worded enquiries. But Mrs. Manigault evidently had no suspicion that anything was wrong between him and Catherine. Her manner towards him was as cordial as always, with the slight effusiveness that subtly hinted that she, a cultivated woman, was aware of his distinction and felt herself honored to have him in her house. Following a custom of long standing between them he bent his head and kissed her hand.

She led the way through a wide hall to a flag-stoned terrace at the back of the house. Chapman recognized the young man in dusty riding clothes who was standing at one end of the terrace, looking out over the fields, as Tom Manigault, although he had not seen him since he was adolescent. He strode towards him. "How are you, Tom?" he said. "I hear you've turned farmer."

Tom did not smile as they shook hands. His eyes met Chapman's with a gaze so direct, so intense that it seemed almost challenging. Chapman, as he accepted a cup of tea from his hostess and sat down on a long cane chair, remembered that Tom, as a boy, had had those remarkable eyes, very much like his mother's. His manner, too, had not changed. At fifteen, he had been so diffident as almost to appear sullen.

"I heard Roy Miller was down here, too," Chapman said, "writing a book. Is he still with you?"

"He left last Tuesday," Mrs. Manigault said in her clear, incisive voice.

"Had he finished his book?"

She shook her head, smiling a little. "I don't think Roy will ever finish that book," she said. "He's been on it five years, at least."

"Well, that's not long, for a book of its scope. He outlined it for me once, at a cocktail party. He was going to start with the Babylonians and come on down to Frank Lloyd Wright." He smiled around the group, in the deprecatory way in which he had come of late to smile whenever dedication to any idea, to any task of intellectual magnitude, was mentioned. "No, that's not long, if he was going to do the job up brown. And Roy has some ambitious ideas."

Mrs. Manigault smiled again. He was startled to observe the fine network of lines etched in a skin that, even when she was in her forties, had seemed to him to have the transparence of a rose leaf. She had been one of the handsomest women he had ever seen. But she was sadly fallen off in the last few years, or was it merely that a change of attitude will suddenly endow a face, a personality, with a new aspect? There was something harassed and even persecuted in the look that she now bent on Catherine.

"Is your grandmother any better today?" she was asking.

Catherine, who had not sat down, but remained poised on the arm of a chair, as if to emphasize the fact that the visit must be brief, shook her head. "We had a pretty bad time with her last night," she said. "She got quite excited, just when it was time for her to go to bed." She raised her head and stared past them all, out over the end of the terrace, to the open fields. "I'll be glad when Willy gets back," she said. "Cousin Daphne irritates her. I think I do too."

"It would drive me nuts to have Cousin Daphne around, even if I was all right to start with," Tom Manigault said. He got up and, lurching with his horseman's stride across the flagstones, set his cup down on a table and stood beside his mother's chair. His thumbs went down, hooked themselves in his belt. Chapman, who was used to finding all the other men in a room smaller than himself, observed the swell of the shoulders under the faded shirt and realized that he had grown into a powerful man. He will be off to war soon, too, he thought, with a recrudescence of the mood that had swayed him earlier in the afternoon. He is too old, too earthy for an air-man, but he will make a splendid captain of artillery. Be killed, too, probably, he thought. He has that look.

Tom spoke with an odd gruffness. "I've got to get back to the field . . . Catherine, how about you all coming over tonight, after you get the old lady in bed?"

Catherine, still poised on the arm of the chair, the empty cup slanting in her hand, seemed to refuse to be wakened out of some dream. She did not turn her head or withdraw her eyes from the fields as she said quietly, "I don't know, Tom . . . I'll see . . ."

Chapman realized that his hostess had bent towards him with some remark. He set a mechanical smile on his face but did not attend what she was saying. He was watching his wife. There was a clairvoyance in her absent look. She seemed, alone and conscious of her loneliness, to be braced against some unimaginable disaster. Something about the pose of the taut, slight figure touched him. The constraint that he had felt in her presence had disappeared. If there were nobody else in the room he would be able to put into words the emotions, the thoughts that had seemed incommunicable.

His excitement grew, manifested itself physically. He became embarrassed for fear some of the company might direct a glance towards him and was relieved when Catherine suddenly rose. "We've got to go," she said. "Thanks, Cousin Elsie. It was good of you to let us stop by."

She was already moving towards the door. He made his good-byes and followed her. Tom Manigault might have been thought to be offended by Catherine's offhand reception of his invitation. He did not address any remark to her and when Chapman approached him he made it convenient not to shake hands with him but merely said a curt "Good-bye."

Outside on the drive Catherine was about to get into the driver's seat but Chapman made a gesture that indicated that she should move over. Heros curled up beside her. Chapman got in and, glad to have some release in physical action of his emotion, relishing, even in his distraction, the feel of the wheel which he had not known now for some weeks, drove at an alarming rate of speed over the rutted road that lay between the two places.

Catherine was silent until they had passed through the farm gate and were descending the slope at the bottom of which the house lay in its grove of trees. "It doesn't change much, does it?" she said then.

"No," he said, "it doesn't change much."

He halted his car in the back door yard, took his bag out of the rumble, and they walked around to the front of the house. Two women were sitting on the porch: old Mrs. Lewis and her cousin, Daphne Passavant. Miss Passavant—she had been briefly married at some time, he seemed to remember, but after she got rid of her husband took back her maiden name—came forward to greet him but the old lady sat, unmoving, her face, on which a faint smile played, a little bent down.

He set his bag on the floor and went over and shook the bony hand, which after a few second's hesitation, she held out to him. In the flowing, immaculate, white linen garments of an old-fashioned cut in which he always envisioned her, she did not seem as much changed as he had expected. Her face, a little heavy in the jowl but remarkable for finely cut features, was still well fleshed. She was still unusually erect for an old woman.

He had not prepared himself for the meeting, but he felt that he must not let his manner towards her show any awareness of her condition, so, drawing a chair a little closer—he knew that she had grown deaf with advancing years—he sat down, saying, "Well, Mrs. Lewis, this is the first time I've ever been here in the fall. You always said I wouldn't really know Swan Quarter till I'd been here in October."

She raised her head. Her faded blue eyes met his, then she bent her head quickly and began toying with the handle of a palm leaf fan that she held in her lap. The touch of the handle, which she kept firmly grasped in her hand while she slowly swayed the fan this way and that, seemed to set going some rhythm which his advent, his abrupt words had shattered. She gazed past him and murmured, in a voice gentler than had been her wont, some words of which he could distinguish only a few phrases. "Two women . . . be glad . . . only for the night . . ."

Cousin Daphne spoke quickly. "She's back in Civil War times."

"What did she say?" Chapman asked in a low voice.

She emitted a sound that was almost a giggle. "She says her father is away, in the service of the Confederacy. We are two women alone. But she sees that you are a gentleman and she'll be glad to have you spend the night."

Chapman controlled the involuntary rictus with which people are apt to greet any sudden manifestation of madness. "I'll be very glad to stay," he said matter-of-factly. "Very glad, indeed." Slightly turning in his chair he elevated his shoulders, shaking his whole big frame, like a dog coming out of water. He looked about him. "How cool it is here," he said. "The train was too hot, all the way down."

Mrs. Lewis gave him another, quick shy glance and murmured something that sounded like "They went away . . ."

Weary as he was and still a prey to violent emotion, he found it difficult to sustain the fiction that he was having a conversation with his hostess. He glanced about for Catherine, but she had disappeared into the house. Cousin Daphne's black eyes were on him. She interpreted swiftly.

"She says the negroes all went away. She talks about that a good deal. She said then that she hoped the Yankees got Ezra . . . She saw him hang a poor, stray cat." She looked away meditatively. "Ezra . . ." she repeated. "I seem to remember hearing about some negro named Ezra who gave a lot of trouble. It was all before my time, of course." She looked back at him. Her mirth was positively indecent now. It shone in two scintillating points in her eyes, made the corners of her mouth twitch helplessly. "The other day Catherine came down in riding clothes," she said, "and *she*" —she stressed the pronoun: it was evident that the old woman had for her ceased to have a personality, would, from now on, be designated in some such fashion—"*she* came up to her and whispered that she didn't blame her. With the country so unsettled and all, she said she'd disguise herself as a man, too, if she had to go travelling. And she tried to give Catherine some money, to help her on her way, she said."

Catherine was standing in the doorway. "Jim, do you want to wash up before supper?" she asked in a cool, hard voice.

He said that he did and took up his bag and followed her up the stairs into the dark, upper hall. She did not enter the room which they had always shared when they were here together but walked farther down the hall and opened the door of the room which Jack Lewis had occupied when he was alive. Her husband followed her inside. He set his bag down on the floor, flung his hat on the bed. She had not advanced very far into the room but still stood at the door. He realized that it was intended that he should occupy the room alone. In a second she would make some hostess-like remark about closets or fresh towels and leave him. He went towards her but there was something at once aloof and forbidding in her expression. He stopped. "Aren't we going to stay in the same room?" he asked in a low voice.

She shook her head, with a quick, agonized motion, and turned and looked out of the window behind her. It was already growing dark but even in the dim light he could discern on her face the same expression she had had a half hour ago, when, poised on the arm of the chair, taking no part in the conversation, she had sat and gazed silently out over the fields. He felt that he could not wait any longer to know what she was thinking. He went to her and took her hand and, putting his other arm about her shoulders, attempted to draw her to him. *"Catherine!"* he said.

She turned her face to his. In the fading light her eyes glittered. With a violent movement she broke away from him. She had reached the window, and, putting her hands behind her, gripped the low sill.

He approached her again, but more cautiously. "Catherine," he said in a low voice, "aren't you ever going to forgive me?"

Her eyes, which he now saw to be swimming with tears, had been on his face. They detached themselves, seemed to fix a point just beyond his shoulder. She spoke, in a low, controlled voice, that was yet shot through with harsh mirth. "I've been living with Tom Manigault for three weeks," she said. "Do you want me back now?"

He felt himself shaken, like a jack in the box, by the spring that a careless hand has, with a touch, released. He knew that he must not go to her, that he must not, on any account, touch her. He forced himself to stand still but all the time he could feel, through the hard soles of his shoes, his feet, seeking the floor that they must cling to, that they must not quit. He spoke quietly, "Will you say that again?"

She flung her head up. "No!" she cried in a harsh voice. "It's none of your business."

He slowly advanced upon her. It seemed a long way. He had had time to change his mind before he set his hands on her throat. He had time to change his mind but he would not change. Standing as close beside her as any lover, he set his hands, the fingers slightly spread, on her throat. He heard the quick, watery in-take of her breath, and then there was no sound in the room except his own breathing, rushing in his ears like a mountain torrent. He had taken hold of her because he wanted only one thing, not to be alone in the abyss into which her words had plunged him. He must see on that smooth face, still slightly wet with tears, a reflection of his own emotions. But he was not satisfied with the expression that confronted him. Except for that quick in-take of breath she had given no sign of revulsion, of attempt to escape. Her eyes were closed. Her face, sharply uplifted to his, was as expressionless as a doll's, except for the slow, ugly widening of the mouth. She opened her eyes now and looked at him. Some part of the horror he had desired to see was in her look. Bending forward, he shook the slight body slowly from side to side, while his thumbs sinking deeper, seemed about to tear from the neck, like a coveted fruit, the rounded column of supporting muscles. At the same time he became aware that somebody was tugging at his wrists, ineffectually but persistently. The hand fell away for a second, then began again its feeble plucking. The rhythm was suddenly broken by a new note. Fingers dug into the cords of his wrists, so sharply that he felt his pulse dulling under the pressure. "Stop!" they seemed to say. "There is something we must tell you," and like the man who, proceeding into the street even when a tugging at his coat sleeve indicates that it is not safe for crossing, will yet halt at the sudden, desperate tug that says "You fool!" so, at the moment when the hand seemed about to fall away for the last time, he slowly relaxed his grip and, as she drew herself upright, staggering slightly, he raised his head and stared about the room, wonderingly, as if asking himself how he had come there and for what purpose.

XXI

SHE HAD BEEN GONE for several minutes before he roused himself. It was now almost dark in the room. A quiet room, furnished primly with a few heavy pieces of old-fashioned furniture, suited to the occupancy of an old maid or an invalid widow. A square of embroidery, framed in walnut, hung over the mantel: poppies, interspersed with sheaves of wheat, executed in faded silks. On the white coverlet of the bed a round, dark object lay, as incongruous as a stone that, flung into a pond, starts no ripples because it has been miraculously suspended on the bosom of the quiet water.

He took his hat up from the bed and, closing the door behind him, walked down the hall. As he approached the head of the stairs his footfalls grew softer. He paused, his hand on the rail, to peer down into the blackness below. The house was filled with an immense quiet. The people who had inhabited it were all gone away or sitting in remote rooms, tranced in silence. He descended the stairs, flinching at the sound of each of his own footfalls, and walked through the lower hall out on to the porch.

Outside, the dark that filled the house was dispersed into a soft, airy grayness. He went down the brick walk, between the crowding box bushes, the crape myrtle, and stepped over the broken fence into a thicket of small growth. The saplings were mostly dogwood or sassafras trees, which, long neglected, had grown to unusual height. Rabbits or some other small animal had made run-ways through the buck-berry bushes that grew up about their trunks. He seemed to remember a path that had once led from this side of the house up past the old graveyard, and he followed the largest of the trails out of the thicket into the deep woods.

He knew an immense relief at getting out of the house. It had seemed to him as he walked softly through the halls that he was not, for all its emptiness, unobserved. Behind each of the closed doors, an occupant, sitting in some attitude of desolation or pacing the floor with hopeless, reiterated gestures had turned to stare bright-eyed after him, who was leaving. "An old, unhappy house," he thought, "a prison, but no part of me. No part of me," he repeated, staring, surprised, between the dark plumes of a poplar up into a sky where a little brightness still lingered. "I do not belong anywhere. There is no place anywhere that is a part of me."

The path ascended. The graveyard lay up that way and above it the big spring. He remembered another, smaller spring down near the creek from which he used to drink when he shot squirrels in these woods. The path branched. He took the turn to the right. He could hear the creek flowing. In some places in this country, where the trees had all been cut, the streams had grown sluggish and yellow, but this patch of woodland had never been cut over or only a little, when Jack Lewis, pressed for money, would sell some of the white oaks. The creek, fed by numerous springs, flowed shallow and clear over wide ledges of limestone.

The path descended sharply, emerged on the banks of the creek. He blundered about among the trees until he found the spring, stretched himself on the ground, drank thirstily of the cold, sweet water and sought the bank of the stream. He stood for a few minutes, staring at the water which still glittered a little in the dying light, then sat down on the arched roots of a sycamore.

Now that he was motionless the horror came back. He reviewed the events of the last hour, like the man who, snatched back from a precipice, collapses on the ground for a moment, then tip-tocs back and peers over the edge, fathoming the depths of the abyss in order to convince himself that he is safe. He had advanced upon her. His hands, which in that second had seemed to have an independent, rebellious life of their own, had seized upon her throat. He had been under a spell and yet he had known what he was doing. There had been a moment, when, advancing upon her, he had reminded himself that what he was about to do carried penalties of various sorts, had asked himself whether it was worth the risk. His greedy hands had said, "Yes." There was only one way in which what she had just said could be erased from consciousness, could be made not to be. It had been a delight to let his hands exert their full strength upon the frail throat. It was her own, agonized, clutching hands that had recalled him to himself. He had gone too far to turn back of himself. He would have gone on. He would have killed her.

He shivered and for a moment cradled his head in his arms. When he raised his head the darkness had increased. The masses of trees that fringed the bank were almost black. On his left, beyond a stretch of sandy beach, there was one low mass of foliage that, rounding out into the water, narrowed the creek to the width of a brook. The opposite bank held no growth; a limestone cliff towered above the tallest trees. There would be a pool at its foot, black and deep.

Fear of showing themselves cowards keeps some men from snapping the slack thread that binds them to life. But if a man were already a coward, a murderer? . . . And who would there be to accuse?

The pool crouched, unvisited, at the foot of the cliff. It would welcome the comer, would send its dark waves flying before the thrusting head, and then the waves would ripple back, would close over the head, the arms that, beating downward, would bring up more and more waves of the dark, to press against the opened lips until the greedy mouth, satiated, would cease to drink the dark, and the lips would close and the whole body would hang motionless or swaying only a little, while the waves closed deeper over it and the pool, gathering its black garment over its unruffled bosom, placidly awaited the coming of the stars.

The stars, he thought, and looked up. But no stars showed in the gray sky, only a thin slice of new moon, glittering palely above the boughs of an oak.

I must go back, he thought. Back to what? And he stood, feeling the dark deepen around him. A shadow detached itself from the low, rounded mass of foliage on the opposite bank, moved forward into the little light that hovered above the sandy beach. Chapman stood, staring at what seemed to be a man's figure, tall but dwarfed by the pack he bore on his back. The shadow wavered an instant, then sank slowly to the sand, as a man, breaking a wearisome journey, might sink at the foot of some tree, casting from him the pack that he had borne all day.

They came this way, Chapman thought. *They came up*

this creek. It was the sycamores led them. They saw sycamores and knew that there was water.

He got up. He went towards the shadowy figure. "It is up here," he said, his voice sounding thin and eery in his own ears. "The spring is up here on the bank, just above that black gum."

Nothing moved in the shadows. Chapman went nearer. "But I would advise you not to settle," he said and laughed. "I know who you are," he said. "You are the son of Irish John Lewis. I know what your father did at the manor," he said and saw the kitchen garden, its cabbages stretching away in delicate, green rows from the weathered stone house. *The stout man in the snuff-colored coat stood, his legs spread, his hands in his pockets. A setter bitch was snuffing at his breeches leg. He took the handle of his whip and was gently pushing her away when the gate clicked and the young, red-faced man came down the path. The stout man had thought that he was alone. He gave a start and his hand went to his coat pocket and came out empty. He stooped and, roughing the dog's head, asked the young man what he wanted and was rising on spraddling legs when the young man, not speaking, except for the low, moaning grunt that broke from him, stooped, too, and struck the exposed, bald head with the shining instrument that he held half-concealed in his hand. The old man had not moved from where he had fallen among the cabbages. There was nothing about him moved, except the blood that trickled down the side of his bald head. But the young man was running. Halfway to the fence he stopped and looked back at the prone figure, then turned the full stare of his shining, insolent eyes on the watcher, and smiled and leaped the fence.*

Chapman felt his hand go out in a gesture of dissent. "It is not that," he said. "I know that he has put the ocean between him and the manor. The sheriff will never find him here. . . . And *you* have fought since then in another cause. At Point Pleasant," he said, and saw on the other side of the stream the painted Wyandot faces waver forward, then fade before the shrill keening of a silver whistle. "A brave man, that man with the silver whistle," he muttered, "and even one-armed the best swordsman in Europe. I wish they had not let the wolves tear his body. . . . At Point Pleasant," he said. "At King's Mountain. And now they have given you this land. Forty degrees from yon blasted oak, three hundred from where the creek, making its bend, flows south. You have not taken the wrong turn. You have read the figures right. This is the land named in your grant. But," he added persuasively, "I would advise you not to settle here. Indeed," he said, going a step nearer, "I would advise you not to settle on this land."

The man in buckskin seemed to rise. He undid his pack and took some articles from it and ranged them on the sand, then lifting the buffalo hide in which they were wrapped, shook it out and hung it on the bough of a tree.

"It is a land of amazing fertility," Chapman said, "well watered and full of game. The pool below us is alive with swans. A mile from here there is a great lick where all the ground shines with salt. . . ."

The man seemed to stoop and in the deep shade to be piling dry branches criss-cross, one upon the other, and powder flashed in the pan and the pure flame rose and wavered against the dark, then burned steadily. The man went to the creek bed and sought a flat stone and laid it at the base of the flame and when glowing coals had ranged themselves about it and it was smoking, laid the bloody tongue of a buffalo upon it.

Chapman waited until he had seated himself again, then spoke. "The land is cursed. It is an old land, ruled by a goddess whose limbs were weary with turning before ever Ireland rose from the sea. An ancient goddess whom men have wakened from an evil dream. . . ."

The man, while he slowly raised a morsel of roasted flesh to his lips, fixed his eyes upon him. "I could wish you had more learning," Chapman said in a courteous tone, "for then I might cite instances from the past that would convince you. I cannot even describe the nature of the enchantment, for it is an enchantment such as never before fell on any people, and how," he asked wearily, "how can I describe the demons who serve the goddess, for you who know only the old-world enchantments?"

He thought that the man stared at him through the leaping flames. It seemed to Chapman that there was mockery in his steady gaze. He pressed closer to the firelight. "It is No Man's Land," he said. "That is the enchantment. The land will turn brittle and fall away from under your children's feet, they will have no fixed habitation, will hold no one spot dearer than another, will roam as savage as the buffalo that now flees your arquebus. And the demons!" he cried. "They will be guarded and served by demons. But the demons will not have the grace, as in ages past, to assume half-human forms, but will retain their own inhuman shapes! Sticks, stones, pulleys, levers such as Archimedes devised, will have voices, show frowning visages. And your children, languid, pale, their members withered, will bow down before them and when they speak their voices will not be heard for the inhuman babble of their gods. I tell you," he cried and heard the thin, minatory shriek rise and float above the quiet trees, "I tell you this tree, whose boughs, gleaming, led you to this stream, felled, its heart reft out, will have a voice, will utter cries louder than any you can utter. Cleena," he cried, and hearing the name, old and forgotten before he came into the world, stretched his hands out to the dark water: "Cleena, he will not listen!"

But the man in buckskin had laid his morsel of flesh down and was rising. He put out his hand. The pack was on his back. He moved, rifle in hand, towards the stream.

The waves parted before the giant, striding limbs and rushed, moaning and gurgling, against the roots of the sycamore. He would reach the bank, start up the path. There would be the great wind of his passing. He might turn his head. There would be the bold, shining stare of his eyes. Chapman got up, held his hands before his face, and stumbled off through the trees.

Afterwards, when he walked in those woods, he could never find the place where he spent the rest of that night: a hollow, drifted thick with dank leaves, between two poplars. Stumbling, as if he were drunk, he had several times fallen to the ground. Once he did not get up, but lay where he fell. The leaves under his spread fingers felt chill and wet. He thought of his ghostly companion, who would have lain down, unthinking, in such a spot, and smiled and composed his limbs and slept. When he woke the sky showed pale above his head. He got up, wandered among the trees until he found the path and went between the hovering box bushes up the brick walk to the house. It was blacker here. Day might not be coming. Something moved beside one of the pillars of the porch. He stopped. "Who is it?" he asked in a low voice.

"Catherine," a voice said.

She sat on the steps, in her dark robe almost invisible among the shadows. He sat down beside her. She did not move or show any awareness of his presence, but he felt a faint, anticipatory shiver of the nerves. The sudden proximity of another human being disturbed him. For a long time now he had seemed to be striding across a gray, uninhabited plain. He strode steadily, almost contentedly, but feeling the need of haste. At any moment the curving, yellowed grasses, the contorted bushes might find a voice and speak. He could stride on past their desolate rustlings, but a human voice would arrest his progress. In a minute she would stir, would say something. There was not much to be lost now, but he would have to listen. If he were not so weary he would get up now and leave the house. It would take effort. He would have to go back into the hall, ring the bell on the old-fashioned wall telephone, lift the receiver from the hook. The operator would be hard to rouse. The bell would peal again and again through the silent house.

He put his hand out. It fell on a round object, woollen to the touch: a man's hat.

"Has somebody been here?" he asked.

"Tom Manigault," she said.

"What did he want?" he asked incuriously.

"He came to find out what was going on between us."

He laughed. "Did you tell him?"

"No," she said, and after a moment laughed too. She rose. "I'm going to make some coffee," she said. "Will you light the fire for me?"

"Certainly," he said and followed her around the side of the house to the kitchen. The door was latched from inside. He set his shoulder against it and pushed. The latch gave with a brittle click. They entered. The room, except for the gray squares of unshuttered windows, was black. He took his pocket lighter out and standing in the middle of the room held the flame high. Catherine, moving surely in the dark, was bending over a box that stood near the stove. Still stooping, she turned her face up to his. "There isn't any wood," she said.

"There must be some," he told her. "Where do they keep it?"

"On the porch, I think. No, in the old out-kitchen. I saw Rodney go in there with a load the other day."

He handed her the lighter and went out the back door and crossed the unfloored brick porch that lay between the two kitchens. The pale morning glimmering through latticed walls revealed lengths of stove wood piled high in a corner. In a deep box beside the pile he found kindling, dry brittle shingles from the roof of the old smokehouse that had tumbled down this spring. He carried an armful of wood into the kitchen, then went back and got an armful of shingles. He lifted the heavy iron lids, the hour-glass-shaped middle section from the top of the stove and, crushing a newspaper that Catherine brought him, carefully arranged his shingles over it, then put the lids back in place.

Catherine was standing beside him. "You do something with the damper," she said.

"I know," he said and shot the lever that would apply a draft to the flame. "I'll put some wood on in a minute," he told her, "but you can set your coffee pot on now."

She had filled the big porcelain pot with water from the cedar bucket and now, with a great iron spoon, measured four tablespoons of coffee into the pot.

He pushed the pot aside, lifted the lid. The flames had risen. He laid three sticks of wood on top of the crackling shingles and replaced the lids.

"It is not much trouble to build a fire in one of these stoves," he said and found a towel and wiped his hands.

She was sitting on the edge of the table, the skirts of her dark robe wrapped tightly about her legs. She smiled faintly. "You always were a good fireman," she said.

He reflected that, unlike most scholars, he had always been ready with his hands. He regretted that the task which for several minutes had occupied all his energies, was finished. "Is there anything else I can do?" he asked politely.

She shook her head. "Maria does something with egg shells, but I'm not going to bother about that."

He stood, listening to the crackling inside the stove. "These old-fashioned pots make excellent coffee," he said.

"Yes, if you take the time."

Time, he thought, but what else have we got? It is mounting in a great wave. It will bear down on us. It will crush us, and he remembered a day at the beach when they had waded out past the ropes and had stood, waiting for the great, mounting swells to lift them from their feet and rock them like sea gulls on the waves. She had never been a strong swimmer and was always reluctant to bathe in the surf, but he knew that each swell that, slowly mounting over their heads, enveloped them in curling green foam, would, when it ebbed, leave them standing together, their feet firmly planted on the tough, slippery seaweed, and he would take her hand and compel her to stand beside him. He would not be capable of even that mild exertion now. He had never felt such deathly weariness.

He went to the door and stood looking out into the gray. Light slid past his shoulder, fell in a yellow oblong on the bricks of the passageway. Catherine had lighted a lamp. The room was fragrant with the smell of the boiling coffee. She came and silently put a cup into his hand. He drank it and turned back into the kitchen. Catherine was standing beside the stove. She set her empty cup down and held her hands out over it. Her hands had always been thin, but now, as she spread them over the glowing lids, the tendons that threaded the fingers stood out, glistening a little in the lamplight. They were no longer the hands of a young girl, but of a frail woman, hovering for warmth over unseen, imprisoned flames.

"You ought to go to bed," he said quietly. "You must be tired."

She raised her head, stared at the window. "Somebody is coming," she said.

They went to the door. Wheels sounded on the road above. They walked around the end of the out-kitchen and stood in the gray light, waiting. The long, black van descended the slope swiftly, stopped in the door yard. A negro boy got out, stood for a second beside the step, then moved, slouching, to the rear of the van. Willy's head appeared. She took the hand that Chapman proffered and stepped down on to the grass.

Catherine moved forward and embraced her. "What made you come this time of night?" she asked.

Without answering, Willy detached herself from the embrace. She turned to the boy, who stood, his arms folded across his chest, leaning against the closed back door of the van. "You wait a minute, Ananias," she said.

She stooped and would have lifted the bag that rested on the grass, but Chapman took it from her. She moved before them towards the house. "What is the matter?" Catherine asked. "Aunt Willy, what is the matter?" She stopped and looked back at the immobile boy. She screamed. *"Red!"* she cried. "Something has happened to Red!"

Willy, moving slowly towards the lighted door, did not turn her head.

"Wait," Chapman said to his wife. He went forward swiftly, came abreast of the slowly moving figure. "Miss Willy, is the horse dead?"

She had achieved the doorway and now she sat down on a stool that was pushed a little way out from the kitchen table.

"He was electrocuted," she said and raised her hand and pushed her hat from her forehead. It fell to the floor. She bent and then, remindful of their presence, straightened up. "There was a loose board on his stall. He kept pushing his head through. And the electric light was dangling there."

"Light?" Chapman repeated, incredulous.

Her pale lips assumed the outline of a smile. "His feet were wet from the manure, you see. They said he just champed it once." Her hands came out in a stiff, hieratic gesture, her whole body slumped downward on the stool. "He was back on all fours, like a dog," she said.

"Like a dog!" Catherine whispered and wailed again. "*Red!* Oh, Red!"

Willy got up from the stool. The smile widened until it seemed carved permanently on her pale face. "We ought to have fixed that board," she said. "We ought to have fixed it that night." She moved into the middle of the room. In those few minutes day had broken. The morning light, flooding the room, made the lamplight pallid. Suddenly, from the yard, a rooster crowed.

Willy went through the door. Outside, on the porch, she bent over a sack of grain. She straightened up, the tin scoop flashing in her hand. She had taken a bucket from the table and was pouring grain into it. She looked up abstractedly. "I'll just go on and feed the chickens," she said. "We'll bury him, as soon as it's light."

"Don't!" Catherine cried. "Oh, Aunt Willy, come to bed."

But Chapman took her by the shoulders and made her stand quiet beside him. "Let her alone," he said.

They heard the screen door close, heard her step down on the ground, heard her speak once to the boy, and then her footfalls died away. Through the open back door they could see the van jutting out past the out-kitchen. Chapman, standing alone now, raised his eyes to the woods. The oak boughs, still green, emerged from the mist in airy shapes. In the door yard a honeysuckle vine, sagging on a rotten trellis, appeared to have every leaf fresh-washed in dew. Something bright flickered out from under its twisted roots. A garden snake, its coat as tender, as bright a green as when it had been spawned in summer heat, ran over the grass to hide in the dark leaves of the quince bush. He looked again at the van where the proud form lay, so long, so still. Arrested by a slight sound, he turned. His wife had blown the lamp out and sat on the end of the table. Her head was bent. She held her hands before her face. The sound he had heard was the slipper falling from her dangling foot. He stooped and was about to slide it back on her foot, when, still holding it in his hand, he bent lower and set his lips on her bare instep.

"Come," he said and heard all the echoes stir in the sleeping house. "We will bury him, as soon as it's light. Then we must go."

CAROLINE GORDON (1895–1981), was a native of Todd County, Kentucky, the locale of much of her fiction. She began her literary career in the 1920's after her marriage to the poet Allen Tate. Her first short story appeared in 1929. *Penhally*, her first published novel, appeared in 1931. It was followed by eight other novels, including *Aleck Maury, Sportsman* (1934), *The Garden of Adonis* (1937), *None Shall Look Back* (1937), *Green Centuries* (1941), *The Strange Children* (1951), *The Malefactors* (1956) and *The Glory of Hera* (1972). *The Women on the Porch* was published in 1944. She published two collections of short stories, *The Forest of the South* (1945) and *Old Red and Other Stories* (1963). She was also the author of *How to Read a Novel* (1957) and, with Tate, of *The House of Fiction: an Anthology of the Short Story* (1950).

LOUISE COWAN holds the doctorate from Vanderbilt University. She was a member of the English faculty and Dean at the University of Dallas, where she was a colleague of Caroline Gordon. She is the author of critical essays on Gordon's fiction, as well as *The Fugitive Group* and *The Southern Critics*. She was awarded the 1992 Charles Frankel Prize for her work in the humanities.